BIOTECHNOLOGY
RESEARCH
IN AN AGE OF TERRORISM

Committee on Research Standards and Practices to Prevent the
Destructive Application of Biotechnology

Development, Security, and Cooperation
Policy and Global Affairs

NATIONAL RESEARCH COUNCIL
OF THE NATIONAL ACADEMIES

THE NATIONAL ACADEMIES PRESS
Washington, D.C.
www.nap.edu

THE NATIONAL ACADEMIES PRESS   500 Fifth Street, N.W.   Washington, DC 20001

NOTICE: The project that is the subject of this report was approved by the Governing Board of the National Research Council, whose members are drawn from the councils of the National Academy of Sciences, the National Academy of Engineering, and the Institute of Medicine. The members of the Committee responsible for the report were chosen for their special competences and with regard for appropriate balance.

The development of this report was supported by the Alfred P. Sloan Foundation and the Nuclear Threat Initiative. Any opinions, findings, conclusions, or recommendations expressed in this publication are those of the author(s) and do not necessarily reflect the views of the organizations or agencies that provided support for the project.

International Standard Book Number 0-309-08977-8 (Book)
International Standard Book Number 0-309-52613-2 (PDF)
Library of Congress Control Number 2003115100

Additional copies of this report are available from The National Academies Press, 2101 Constitution Avenue, N.W., Lockbox 285, Washington, DC 20055; (800) 624-6242 or (202) 334-3313 (in the Washington metropolitan area); Internet, www.nap.edu

Printed in the United States of America

THE NATIONAL ACADEMIES
Advisers to the Nation on Science, Engineering, and Medicine

The **National Academy of Sciences** is a private, nonprofit, self-perpetuating society of distinguished scholars engaged in scientific and engineering research, dedicated to the furtherance of science and technology and to their use for the general welfare. Upon the authority of the charter granted to it by the Congress in 1863, the Academy has a mandate that requires it to advise the federal government on scientific and technical matters. Dr. Bruce M. Alberts is president of the National Academy of Sciences.

The **National Academy of Engineering** was established in 1964, under the charter of the National Academy of Sciences, as a parallel organization of outstanding engineers. It is autonomous in its administration and in the selection of its members, sharing with the National Academy of Sciences the responsibility for advising the federal government. The National Academy of Engineering also sponsors engineering programs aimed at meeting national needs, encourages education and research, and recognizes the superior achievements of engineers. Dr. Wm. A. Wulf is president of the National Academy of Engineering.

The **Institute of Medicine** was established in 1970 by the National Academy of Sciences to secure the services of eminent members of appropriate professions in the examination of policy matters pertaining to the health of the public. The Institute acts under the responsibility given to the National Academy of Sciences by its congressional charter to be an adviser to the federal government and, upon its own initiative, to identify issues of medical care, research, and education. Dr. Harvey V. Fineberg is president of the Institute of Medicine.

The **National Research Council** was organized by the National Academy of Sciences in 1916 to associate the broad community of science and technology with the Academy's purposes of furthering knowledge and advising the federal government. Functioning in accordance with general policies determined by the Academy, the Council has become the principal operating agency of both the National Academy of Sciences and the National Academy of Engineering in providing services to the government, the public, and the scientific and engineering communities. The Council is administered jointly by both Academies and the Institute of Medicine. Dr. Bruce M. Alberts and Dr. Wm. A. Wulf are chair and vice chair, respectively, of the National Research Council.

www.national-academies.org

Preface

The charge to our Committee was to consider ways to minimize threats from biological warfare and bioterrorism without hindering the progress of biotechnology, which is essential for the health of the nation. This task is complicated because almost all biotechnology in service of human health can be subverted for misuse by hostile individuals or nations. The major vehicles of bioterrorism, at least in the near term, are likely to be based on materials and techniques that are available throughout the world and are easily acquired. Most importantly, a critical element of our defense against bioterrorism is the accelerated development of biotechnology to advance our ability to detect and cure disease. Since the development of biotechnology is facilitated by the sharing of ideas and materials, open communication offers the best security against bioterrorism. The tension between the spread of technologies that protect us and the spread of technologies that threaten us is the crux of the dilemma.

Although the National Academies have had many reports on national security, this is the first to deal specifically with national security and the life sciences. The thoughtful report on *Scientific Communication and National Security* (National Academy Press, 1982) had as its charge "to examine the relation between scientific communication and national security in light of the growing concern that foreign nations are gaining military advantage from such research"; however, it did not deal with the life sciences. Since that report, much has happened to justify an examination of the life sciences in this context—the discovery of nations with clandestine research programs dedicated to the creation of biological weapons, the anthrax attacks of 2001, the rapid pace of progress in biotechnology, and the accessibility of these new technologies via the Internet. All of these developments have prompted the current report. The goal of this report is to make recommendations that achieve an appropriate balance between the pursuit of scientific advances to improve human health and welfare and national security.

In preparing this report our Committee examined ways by which the spread of technology, methods, materials and information could be limited to constructive activities concerned with medical progress. The dual use nature of these advances strongly argues that any initiative must demonstrably increase our net security. Erring on the side of prudence and favoring the inhibition of information flow could retard the development of successful defenses and seriously compromise our nation's health. Therefore, the challenge is for the scientific community to develop a system that permits fundamental research to proceed unimpeded, while identifying research with great potential for misuse.

The scientific community historically has demonstrated its ability to lead the way in the responsible development of new technologies. After the Asilomar conference in 1975, scientists designed and followed a set of guidelines for work with recombinant DNA, then a novel technology of unexplored potential. These guidelines, keyed to the risk of exposure to genetically modified organisms, have prevented any untoward events, reassured the public, and allowed the rapid and efficient progress of academic and commercial applications of these technologies. The recombinant DNA guidelines were established to prevent unintended creation of harmful recombinant organisms. But now the nation faces a different problem: the intentional use of biotechnology for destructive purposes. This challenge must engage the entire community of biologists nationally and internationally. In a joint statement issued on November 8, 2002, and printed in the journal *Science*, the presidents of the U.S. National Academy of Sciences and the U.K. Royal Society called on scientists to assist their governments in combating the threat of bioterrorism: "Today, researchers in the biological sciences again need to take responsibility for helping to prevent the potential misuses of their work, while being careful to preserve the vitality of their disciplines as required to contribute to human welfare."

To consider ways to balance national security and scientific openness, the Committee had six meetings held in Washington, D.C. between April 1, 2002 and January 29, 2003. Representatives from the National Institutes of Health, the Executive Office of the President, governmental and nongovernmental technical and policy experts, and educators and private consultants briefed the Committee. The Committee also reviewed information available from the open literature as well as new materials prepared by experts (see Appendix C).

During the course of our deliberations, Committee members recommended that scientific, policymaking, and intelligence communities be brought together to focus on the challenges raised by advances in biotechnology. To this end the National Academies and the Center for Strategic and International Studies convened a one-day meeting on "Scientific Openness and National Security" in Washington, D.C., on January 9, 2003. This meeting emphasized the importance of a continuing dialogue between the life sciences and the intelligence communities both nationally and internationally. It is our hope that this report provides the basis for this dialogue.

The Committee wishes to express its sincere appreciation to the devoted project staff. As study director, Eileen Choffnes ensured the success of this project through her expertise, dedication, and creativity. This study would not have been possible without Dr. Choffnes' oversight and coordination of the work of the Committee and her insightful editing of the report. Amy Giamis was outstanding in her great finesse in the organizational work of the Committee, and her numerous contributions to supporting the research and editing of the report. Finally, the Committee wishes to express its appreciation to Jo Husbands, who brought to our deliberations considerable insights from her experience as Director of the NAS Committee on International Security and Arms Control. Throughout the study we were encouraged by the support of NAS President Bruce Alberts. I want to express my personal thanks to the individual members of the Committee for the dedication and energy with which they tackled this difficult problem. The report would not have been possible without the perspectives of these experts, who represented their diverse disciplines so eloquently.

Gerald R. Fink
Chair

Acknowledgments

This report has been reviewed in draft form by individuals chosen for their diverse perspectives and technical expertise, in accordance with procedures approved by the NRC's Report Review Committee. The purpose of this independent review is to provide candid and critical comments that will assist the institution in making its published report as sound as possible and to ensure that the report meets institutional standards for objectivity, evidence, and responsiveness to the study charge. The review comments and draft manuscript remain confidential to protect the integrity of the deliberative process.

We wish to thank the following individuals for their review of this report: Kenneth Berns, Mount Sinai Medical Center; James Cook, Washington State University; Malcolm Dando, University of Bradford; Catherine DeAngelis, *Journal of the American Medical Association*; Stanley Falkow, Stanford University; Claire Fraser, The Institute for Genomic Research; Michael Friedman, City of Hope; Robert Haselkorn, University of Chicago; Michael McGeary, McGeary and Smith; Michael Moodie, Chemical and Biological Arms Control Institute; Harley Moon, Iowa State University; Stephen S. Morse, Columbia University; Jerome Schultz, University of Pittsburgh; Jonathan Tucker, Center for Nonproliferation Studies, Monterey Institute of International Studies; and Mark Wheelis, University of California, Davis.

Although the reviewers listed above have provided many constructive comments and suggestions, they were not asked to endorse the conclusions or recommendations, nor did they see the final draft of the report before its release. The review of this report was overseen by Gilbert Omenn, University of Michigan Medical School, and R. Stephen Berry, University of Chicago. Appointed by the National Research Council, they were responsible for making certain that an independent examination of this report was carried out in accordance with institutional procedures and that all review comments were carefully considered. Responsibility for the final content of this report rests entirely with the authoring committee and the institution.

Contents

BIOTECHNOLOGY
RESEARCH
IN AN AGE OF TERRORISM

Executive Summary

The great achievements of molecular biology and genetics over the last 50 years have produced advances in agriculture and industrial processes and have revolutionized the practice of medicine. The very technologies that fueled these benefits to society, however, pose a potential risk as well—the possibility that these technologies could also be used to create the next generation of biological weapons. Biotechnology represents a "dual use" dilemma in which the same technologies can be used legitimately for human betterment and misused for bioterrorism.

This report reflects the increasing attention being paid by scientists and policymakers to the potential for misuse of biotechnology by hostile individuals or nations and to the policy proposals that could be applied to minimize or mitigate those threats. The term "misuse of biotechnology" is a phrase that captures a wide spectrum of potentially dangerous activities from spreading common pathogens (e.g., spraying *Salmonella* on salad bars) to sci-fi plots of transforming pathogens into the next "Andromeda strain." Our Committee addressed one important part of this spectrum of risks of potential misuse: the capacity for advanced biological research activities to cause disruption or harm, potentially on a catastrophic scale. Broadly stated, that capacity consists of two elements: (1) the risk that dangerous agents that are the subject of research will be stolen or diverted for malevolent purposes; and (2) the risk that the research results, knowledge, or techniques could facilitate the creation of "novel" pathogens with unique properties or create entirely new classes of threat agents. The charge to the Committee was to:

• Review the current rules, regulations, and institutional arrangements and processes in the United States that provide oversight of research on pathogens and potentially dangerous biotechnology research, within government laboratories, universities and other research institutions, and industry.

• Assess the adequacy of current U.S. rules, regulations, and institutional arrangements and processes to prevent the destructive application of biotechnology research.

• Recommend changes in these practices that could improve U.S. capacity to prevent the destructive application of biotechnology research while still enabling legitimate research to be conducted.

Although the focus of the report is on the United States, this country is only one of many pursuing biotechnology research at the highest level. The techniques, reagents, and information that could be used for offensive purposes are readily available and accessible. Moreover, the expertise and know-how to use or misuse them is distributed across the globe. Without international consensus and consistent guidelines for overseeing research in advanced biotechnology, limitations on certain types of research in the United States would only impede the progress of biomedical research here and undermine our own national interests. It is entirely appropriate for the United States to develop a system to provide oversight of research activities domestically, but the effort will ultimately afford little protection if it is not adopted internationally. This is a challenge for governments, international organizations, and the entire international scientific community. Efforts to meet that challenge are under way, but they must be quickly expanded, strengthened, and harmonized.

THE CURRENT AND EVOLVING REGULATORY ENVIRONMENT

In the United States, the USA PATRIOT Act of 2001 and the Bioterrorism Preparedness and Response Act of 2002 already establish the statutory and regulatory basis for protecting biological materials from inadvertent misuse. Once fully implemented, the mandated registration for possession of certain pathogens (the "select agents"), designation of restricted individuals who may not possess select agents, and a regulatory system for the physical security of the most dangerous pathogens within the United States will provide a useful accounting of domestic laboratories engaged in legitimate research and some reduction in the risk of pathogens acquired from designated facilities falling into the hands of terrorists. The Committee stresses that implementation of current legislation must not be overly restrictive given the critical role that the development of effective vaccines, diagnostics, therapeutics, and detection systems, along with a responsive public health system, will play in providing protection against bioterrorism—and other serious health threats. Otherwise these legislative solutions may unintentionally limit the research on dangerous pathogens by legitimate laboratories and investigators. To be effective, a harmonized international system for the regulatory oversight of the possession of dangerous pathogens and toxins, comparable to the one being put in place in the United States, is needed.

With regard to oversight of research, no country has developed guidelines and practices to address all aspects of biotechnology research. The Committee has concluded that existing domestic and international guidelines and regulations for the conduct of basic or applied genetic engineering research may ensure the physical safety of laboratory workers and the surrounding environment from contact with or exposure to pathogenic agents or "novel" organisms. However, they do not currently address the potential for misuse of the tools, technology, or knowledge base of this research enterprise for offensive military or terrorist purposes. In addition, no national or international review body currently has the legal authority or self-governance responsibility to evaluate a proposed research activity prior to its conduct to determine whether the risks associated with the proposed research, and its potential for misuse, outweigh its potential benefits. The Committee concluded that the existing fragmentary system must be adapted, enhanced, supplemented, and linked to provide a system of oversight that will give confidence that the potential risks of misuse of dual use research are being adequately addressed while enabling vital research to go forward. The significant increases in funding that will be going to research on biodefense —precisely the sort of research likely to pose the most severe dual use dilemmas—reinforce the argument for creating such a comprehensive system, both nationally and internationally.

A PROPOSED NEW SYSTEM

The system the Committee proposes would establish a number of stages at which experiments and eventually their results would be reviewed to provide reassurance that advances in biotechnology with potential applications for bioterrorism or biological weapons development receive responsible oversight. The system relies heavily on a mix of voluntary self-governance by the scientific community and expansion of an existing regulatory process that itself grew out of an earlier response by the scientific community to the perceived risks associated with gene-splicing research. This is the system created to implement the National Institutes of Health Guidelines for Research Involving rDNA Molecules ("the

Guidelines"). We recognize that successfully implementing the system we propose will require significant additional resources at each stage; we do not attempt to provide an estimate of these costs.

Recommendation 1: Educating the Scientific Community
We recommend that national and international professional societies and related organizations and institutions create programs to educate scientists about the nature of the dual use dilemma in biotechnology and their responsibilities to mitigate its risks.

Adequately addressing the potential risks that research in advanced biotechnology could be used by hostile parties will require educating the community of life scientists, both about the nature of these risks and about the responsibilities of scientists to address and to manage them. At present, awareness of the potential for misuse of biological knowledge varies widely in the research community. Researchers currently working with select agents are already taking steps to contain these agents physically and protect against planned or unplanned harm. But most life scientists have had little direct experience with the issues of biological weapons and bioterrorism since the advent of the Biological Weapons Convention in the early 1970s, so these researchers lack the experience and historical precedent of considering the potential for misuse of their discoveries.

We recommend that the professional societies in the life sciences undertake a regular series of meetings and symposia, in the United States and overseas, to provide both knowledge and opportunities for discussion. It could be useful for one of the major professional societies or science policy organizations to convene a meeting of all the major societies to discuss how best to implement such a program. Industry groups and associations of higher education and research could also usefully undertake the education of their members about the risks and their implications for research practices.

Substantive knowledge of the potential risks is not sufficient, however. The Committee believes that biological scientists have an affirmative moral duty to avoid contributing to the advancement of biowarfare or bioterrorism. Individuals are never morally obligated to do the impossible, and so scientists cannot be expected to *ensure* that the knowledge they generate will never assist in advancing biowarfare or bioterrorism. However, scientists can and should take reasonable steps to minimize this possibility. The Committee believes that it is the responsibility of the research community, including scientific societies and organizations, to define what these reasonable steps entail and to provide scientists with the education, skills, and support they need to honor these steps.

Recommendation 2: Review of Plans for Experiments
We recommend that the Department of Health and Human Services (DHHS) augment the already established system for review of experiments involving recombinant DNA conducted by the National Institutes of Health to create a review system for seven classes of experiments (the Experiments of Concern) involving microbial agents that raise concerns about their potential for misuse.

This part of the system includes both the criteria for deciding which experiments will be subject to review and the process by which the review will take place.

The Criteria for Review. The Committee identified seven classes of experiments that it believes illustrate the types of endeavors or discoveries that will require review and discussion by informed members of the scientific and medical community before they are undertaken or, if carried out, before they are published in full detail. They include experiments that:

1. **Would demonstrate how to render a vaccine ineffective.** This would apply to both human and animal vaccines. Creation of a vaccine-resistant smallpox virus would fall into this class of experiments.
2. **Would confer resistance to therapeutically useful antibiotics or antiviral agents.** This would apply to therapeutic agents that are used to control disease agents in humans, animals, or crops. Introduction of ciprofloxacin resistance in *Bacillus anthracis* would fall in this class.
3. **Would enhance the virulence of a pathogen or render a nonpathogen virulent.** This would apply to plant, animal, and human pathogens. Introduction of cereolysin toxin gene into *Bacillus anthracis* would fall into this class.
4. **Would increase transmissibility of a pathogen.** This would include enhancing transmission within or between species. Altering vector competence to enhance disease transmission would also fall into this class.
5. **Would alter the host range of a pathogen.** This would include making nonzoonotics into zoonotic agents. Altering the tropism of viruses would fit into this class.
6. **Would enable the evasion of diagnostic/detection modalities.** This could include microencapsulation to avoid antibody-based detection and/or the alteration of gene sequences to avoid detection by established molecular methods.
7. **Would enable the weaponization of a biological agent or toxin.**

This would include the environmental stabilization of pathogens. Synthesis of smallpox virus would fall into this class of experiments.

These categories represent experiments that are feasible with existing knowledge and technologies or with advances that the Committee could anticipate occurring in the near future. Some of them represent the types of naturally occurring genetic changes in pathogens that have led to disease pandemics such as the "Spanish Flu" in 1917-1918 or the recently recognized disease "severe acute respiratory syndrome" (SARS) but that could now be engineered in the laboratory. Others have been part of the history of biowarfare research and development. The concerns deal with infectious agents or their products because we believe that self-replicating agents or their products pose the most imminent biological threat.

The seven areas of concern address only potential microbial threats. Modern biological research is much broader, encompassing all of the health sciences, agriculture and veterinary science, and a variety of industrial applications. Moreover, all of these areas are changing rapidly. The great diversity as well as the pace of change make it imprudent to project the potential both for good and ill too broadly and too far into the future. Therefore, the Committee has initially limited its concerns to cover those possibilities that represent a plausible danger and has tried to avoid improbable scenarios. Over time, however, the Committee believes it will be necessary to expand the experiments of concern to cover a significantly wider range of potential threats.

The Review Process. The NIH Guidelines require creation of an Institutional Biosafety Committee (IBC) when research is conducted at or sponsored by an entity receiving any NIH support for recombinant DNA research. Most of the 400 or so IBCs registered with NIH are at institutions that are subject to the NIH Guidelines and for whom IBC registration is mandatory. While most of these institutions are academic, some industry-based IBCs are registered with NIH as a consequence of receiving NIH support. In other instances, companies voluntarily comply with the NIH Guidelines as a means of observing a "gold standard" for safety practices. Several federal agencies and laboratories have made compliance with the NIH Guidelines a condition of their support of intramural and extramural research projects. Furthermore, a number of federal IBCs are registered with NIH.

All of the experiments that fall within the seven areas of concern should currently require review by an IBC. The Committee thus recommends relying on the system of IBCs as the first review tier for experiments of concern.

The Committee recommends that the form researchers now use to submit their experimental designs to the IBC be amended to include another category where researchers would designate whether, in their judgment, their proposed projects fall under an area of concern. The IBC would then review that issue along with the other aspects of the project that it is evaluating, carefully weighing potential benefits versus potential danger. Occasionally, the IBC may discover that what is proposed is forbidden under current guidelines and would not approve the research. In most cases, however, it would designate the project either as acceptable to move forward or as raising concerns that need further consideration at a higher level.

The Committee recommends initial review by the IBC because this provides an assessment of research at its earliest stages. By the time the work is submitted for publication, substantial information about the research may have already been disseminated through informal professional contacts or presentations of preliminary results at scientific meetings. These aspects of the open culture in the life sciences emphasize how important it is to make scientists aware of their personal responsibilities to consider the balances of risks and benefits in their proposed research so they can responsibly inform the IBC.

Experiments that need further consideration would be referred to an expanded Recombinant DNA Advisory Committee (RAC) and possibly to the director of the NIH for approval or denial of permission to proceed with the proposed experiment. The Committee recommends this route because so many of the experiments in the areas of concern would fall under the purview of the RAC already and because it has an established track record of facilitating research while protecting public safety. Under our recommendation, the RAC would begin to review some projects in the areas of concern from all relevant research institutions. This would be a substantial expansion from its current jurisdiction over research funded by NIH and those institutions that comply voluntarily.

When the RAC takes up this new duty, the Committee proposes that it initially translate the categories of experiments of concern into a more detailed set of guidelines for IBCs to use. It should then improve and update these guidelines as needed as its experience with the process grows. The RAC will need substantial new resources to take on this additional task, and both it and the IBCs may need to incorporate new expertise to handle the task.

Recommendation 3: Review at the Publication Stage
We recommend relying on self-governance by scientists and scientific journals to review publications for their potential national security risks.

Publication of research results provides the vehicle for the widest dissemination, including to those who would misuse them. The Committee believes strongly that this part of the system should be based on the voluntary self-governance of the scientific community rather than formal regulation by government.

Proposals to limit publication have caused great concern and controversy among both scientists and publishers. The norm of open communication is one of the most powerful in science. To limit the information available in the methods section of journal articles would violate the norm that all experimental results should be open to challenge by others. But not to do so is potentially to provide important information to biowarfare programs in other countries or to terrorist groups.

Ultimately, any process to review publications for their potential national security risks would have to be acceptable to the wide variety of journals in the life sciences, both in the United States and internationally. The Committee believes that continued discussion among those involved in publishing journals—and between editors and the national security community—will be essential to creating a system that is considered responsive to the risks but also credible with the research community. The Committee believes that the statement produced by a group of editors from major life science journals in February 2003 is an important step in this process.

On the broader question of classification, the Committee believes that the principle set out by the Reagan Administration in 1985 in National Security Decision Directive 189—that the results of fundamental research should be unrestricted to the maximum extent possible and that classification should be the mechanism for what control might be required—remains valid and should continue to be the basis for U.S. policy. The Committee's support for self-governance by the scientific community through appropriate reviews by journals and other publication outlets should not be construed as endorsing the creation of "sensitive but unclassified" information in the life sciences. The Committee believes that the risks of a chilling effect on biodefense research vital to U.S. national security as the result of inevitably general and vague categories is at present significantly greater than the risks posed by inadvertent publication of potentially dangerous results. A system of review based in scientific self-governance can, we believe, effectively address the security risks without discouraging scientists from taking part in important biodefense research.

Recommendation 4: Creation of a National Science Advisory Board for Biodefense

We recommend that the Department of Health and Human Services create a National Science Advisory Board for Biodefense (NSABB) to provide advice, guidance, and leadership for the system of review and oversight we are proposing.

The NSABB would serve a number of important functions for both the scientific community and the government.

- At the most general (strategic) level, it would serve as a point of continuing dialogue between the scientific community and the national security community and as a forum for addressing issues of interest or concern. At the operational (tactical) level, it would provide case-specific advice on the oversight of research and the communication and dissemination of life sciences research information that is relevant for national security and biodefense purposes. Because of its important bridging functions, its members should include both leading scientists and national security experts, including those with experience in managing scientific research in federal agencies.

- In terms of the regulatory aspects of the operation of our proposed system, we recommend that the Board periodically review and suggest updates to the "Experiments of Concern." We also recommend that the Board review and suggest updates to the list of "select agents" and to policies regarding the international exchange of biological agents. A review of the select agents list by DHHS is already required every two years but the Board could serve a useful and important function by providing an independent assessment as an input to that process.

- For the system's self-governing phases, we recommend that the NSABB serve as a resource. This could include aiding the professional societies in developing education programs, as well as providing a convening mechanism. It could also include assisting those producing publications in the life sciences. The Board could provide a convening mechanism for journal editors, organizing periodic discussions among them as they develop and evaluate their review processes. The Board could review and comment on proposed procedures on request, and perhaps serve as a clearinghouse so that journals that have not already adopted review procedures could have ready access to examples of what others are doing. It would be very important for the Board to reach out beyond the United States to the many international publications in the life sciences and to find ways to include their leaders in discussions. The Board might also provide advice on request about particular manuscripts that raise concern, perhaps by organizing small groups of experts to assess the trade-offs between the scientific merits of the research, especially that with the potential to advance knowledge relevant to biodefense, and the risks of publishing information that might assist terrorists or proliferant states.

• In addition to its functions related to the potential risks of research in advanced biotechnology, the Board should have the capacity to advise the government on how the life sciences can contribute to alleviating the risks of bioterrorism and biological weapons through new research in areas such as vaccine, antiviral, and antibiotic development, new detection devices and technologies, and preventive public health measures. This advisory function would serve as a continuous reminder that any system of review and oversight must operate in ways that do not put the United States—and the world—at risk of losing the great potential benefits of biotechnology. Having a Board that was informed and aware of the latest research developments, even including manuscripts not yet published, would provide the capacity for "early warning," alerting the government to the risks of new findings or techniques that should be met by focusing research resources on appropriate responses or countermeasures.

As for the organizational location for the NSABB, there are clear trade-offs between an independent board that offers its advice to government and one that is a formal advisory body to one or more federal agencies. No solution meets all the criteria, but on balance we believe that the logical organizational location for the NSABB is within the Department of Health and Human Services providing advice to the secretary of that Department. DHHS already has a leading role in biotechnology research, particularly that related to the Experiments of Concern. Location within the DHHS would also connect the Board directly to the other parts of our proposed system, the RAC and the IBCs, while not limiting its capacity to work with other relevant agencies or private groups.

International coordination and cooperation will be necessary to make any effort to mitigate the risks of bioterrorism effective. Therefore, in the view of the Committee, the establishment of an NSABB within the United States can serve as the basis for international dialogue aimed at reducing the risks of subversion of legitimate life sciences research efforts. Review systems, comparable to the one proposed involving the IBC and RAC, already exist in many nations. These were established as an outgrowth of the Asilomar conference in 1975. In the same manner, other countries should be encouraged to establish counterparts to the NSABB so that the community of life scientists globally can work together to reduce the risks of offensive applications of life sciences research.

Recommendation 5: Additional Elements for Protection Against Misuse
We recommend that the federal government rely on the implementation of current legislation and regulation, with periodic review by the

NSABB, to provide protection of biological materials and supervision of personnel working with these materials.

There are other elements of the current regulatory system that the Committee believes should be reviewed and evaluated because of their important impact on the conduct of research.

Physical Containment. Safeguarding the collections of existing agents is an obvious priority that in large measure is being addressed through recently passed legislation and implementing regulations. The designation of certain pathogens as "select agents" is an appropriate starting point for identifying strains and isolates that need to be secured. It is crucial to avoid well-meaning but counterproductive regulations on pathogens. Rules for containment and registration of potentially dangerous materials must be based on scientific risk assessment and informed by a realistic appraisal of their scientific implications. Moreover, scientific input is essential to ensure that these rules are clear as well as responsive to periodic assessment of the current technologies and capacities. The NSABB could be available to provide advice on short notice about revising regulations in response to new developments. Rules governing transfer of materials between laboratories to prevent unauthorized distribution or diversion might also be regularly reviewed by the NSABB so that new threats could be recognized and responded to and unnecessary impediments identified for removal.

Trained Personnel. In some areas of technology, the limiting ingredient is the existence of trained personnel. General microbiological training sufficient for culturing and growing pathogenic microorganisms at levels of significant concern is available in high school and first-year college biology courses; majors in microbiology would be sophisticated enough to grow many select organisms. Moreover, training in basic microbiology is widely available outside the United States. The procedures for admitting foreign students and scientists to the United States for study and collaborative research must reflect the importance of keeping universities as open educational environments. Efforts to identify or control knowledgeable personnel within the United States are impractical, and surveillance of such personnel would not, in our opinion, offer much security.

Recommendation 6: A Role for the Life Sciences in Efforts to Prevent Bioterrorism and Biowarfare
We recommend that the national security and law enforcement communities develop new channels of sustained communication with the life sciences community about how to mitigate the risks of bioterrorism.

By signing and ratifying the Biological and Toxin Weapons Convention (BWC), the United States renounced the use and possession of such offensive weapons and methods to disseminate and deliver them. Given the increased investments in biodefense research in the United States, it is imperative that the United States conduct its legitimate defensive activities in an open and transparent manner. This should clear the way for all biomedical scientists to contribute to the development of defensive measures that would mitigate the impact of the use of such weapons against people, plants, and animals.

The intelligence and law enforcement agencies need the academic scientists both for the expertise they might provide about the nature of current agents and the potential for new ones and for the best advice on limiting the spread of new technologies that would make countermeasures more difficult. It might be desirable for components of the national security community to establish advisory boards of basic scientists and clinicians with expertise in areas such as viral disease, bacterial pathogens, biotechnology, immunology, toxins, and public health, as well as others in the area of basic molecular biology. These advisory boards could help members of the intelligence and law enforcement communities keep current in relevant areas of science and technology and provide a trusted set of advisors to answer technical questions.

Recommendation 7: Harmonized International Oversight
We recommend that the international policymaking and scientific communities create an International Forum on Biosecurity to develop and promote harmonized national, regional, and international measures that will provide a counterpart to the system we recommend for the United States.

Any serious attempt to reduce the risks associated with biotechnology must ultimately be international in scope, because the technologies that could be misused are available and being developed throughout the globe. A number of countries and regional and international organizations are already moving forward to develop programs and policies on aspects of the problem; the initiatives include consultations among the parties to the BWC on best practices for the security and oversight of pathogens and toxins. These approaches must be harmonized and widely adopted in order for them to be effective. Just as the scientific community in the United States must become deeply and directly engaged, the commitment of the international scientific community to these issues is needed to implement the recommendations contained in this report.

We do not expect our recommendations to provide a "road map" that could simply be adopted internationally without significant modifications or adaptations to local or regional conditions. But any effective system should include all the issues addressed by our recommendations. The Committee therefore recommends, as a next step, convening an "International Forum on Biological Security" to begin a dialogue within and between the life sciences and the policymaking communities internationally. Among the topics for this international forum are:

- Education of the scientific community globally, including curricula, professional symposia, and training programs to raise awareness of potential threats and modalities for reducing risks as well as to highlight ethical issues associated with the conduct of biological science.
- Design of mechanisms for international jurisdiction that would foster cooperation in identifying and apprehending individuals who commit acts of bioterrorism.
- Development of an internationally harmonized regime for control of pathogens within and between laboratories and facilities.
- Development of systems of review to provide oversight of research, including defining an international norm for identifying and managing "experiments of concern."
- Development of an international norm for the dissemination of "sensitive" information in the life sciences.

This and other forums should be sponsored by international organizations with standing and credibility within both the policymaking and scientific communities. Different topics within this broad agenda may be more appropriate for different organizations. Potential sponsors could include the World Health Organization and the United Nations Educational, Scientific and Cultural Organization (UNESCO) as formal international governmental organizations with direct links to government policymakers. Among nongovernmental scientific organizations are the International Council for Science and more recently created organizations of the world's academies of science such as the InterAcademy Panel on International Issues (IAP) and the InterAcademy Council (IAC) that seek to bring the prestige and convening capacity of these bodies to bear on crucial international problems.

CONCLUSION

Throughout the Committee's deliberations there was a concern that policies to counter biological threats should not be so broad as to impinge upon the ability of the life sciences community to continue its role of contributing to the betterment of life and improving defenses against biological threats. Caution must be exercised in adopting policy measures to respond to this threat so that the intended ends will be achieved without creating "unintended consequences." On the other hand, the potential threat from the misuse of current and future biological research is a challenge to which policymakers and the scientific community must respond. The system proposed in this report is intended as a first step in what will be a long and continuously evolving process to maintain an optimal balance of risks and rewards. The Committee believes that building upon processes that are already known and trusted and relying on the capacity of life scientists to develop appropriate mechanisms for self-governance, offers the greatest potential to find the right balance. This system may provide a model for the development of policies in other countries. Only a system of international guidelines and review will ultimately minimize the potential for the misuse of biotechnology.

1
Introduction

The great achievements of molecular biology and genetics over the last 50 years have produced advances in agriculture and industrial processes and have revolutionized the practice of medicine. The very technologies that fueled these benefits to society, however, pose a potential risk as well—the possibility that these technologies could also be used to create the next generation of biological weapons. Biotechnology represents a "dual use" dilemma in which the same technologies can be used legitimately for human betterment and misused for bioterrorism.

Events over the 1990s focused growing attention on this balance of risks and benefits, part of a larger concern about the proliferation of weapons of mass destruction (WMD)—chemical, nuclear, or biological. In early 1992, President Yeltsin acknowledged that, despite being an original signatory and State party to the Biological and Toxin Weapons Convention (BWC), the Soviet Union had maintained a major clandestine biological weapons program into the early 1990s.[1] Yeltsin ordered the program shut down, but concerns about other possible secret programs remained. Policymakers in the United States became increasingly concerned that so-called "rogue states" would turn to WMD to counter the overwhelming U.S. conventional military superiority. Secretary of Defense Les Aspin launched the "Defense Counterproliferation Initiative" in December 1993 to develop additional means to address these threats. Official statements continue to cite at least a dozen countries believed to have or to be pursuing a biological weapons capability.[2] The terrorist attacks of September 11, 2001 and the subsequent anthrax letters accelerated already existing concerns that terrorists would seek WMD capabilities as well. President Bush, in a speech at West Point in

15

2002, said: "The gravest danger to freedom lies at the perilous crossroads of radicalism and technology. When the spread of chemical and biological and nuclear weapons, along with ballistic missile technology—when that occurs, even weak states and small groups could attain a catastrophic power to strike great nations."[3] States, groups, and individuals are pursuing a biological weapons capability—and the means for them to do so are widely available. U.S. and British concerns about Iraq's reported biological and other WMD programs in early 2003 were a primary reason for launching preemptive military action to find and destroy these weapons capabilities.[4]

Biological weapons have long been stigmatized as "indiscriminant agents of unnecessary suffering, [whose] use ... contradict(s) the universal principles of war."[5] As discussed below, since November 1969 the U.S. programs linked to biological weapons have been restricted to research and development on defensive measures only. Thus few biologists in the United States today have knowledge of our country's past offensive weapons programs or of the concerns of the national security branches of government. In this respect the life sciences community is in a different situation from that of the physics community, which in large part has been continuously involved in government-sponsored weapons research programs since at least World War II. The scientific community and the government jointly face a double challenge: (1) to establish a working relationship with the national security branches of government, and (2) to help craft a system that will minimize the risk of wrongful use of biological agents or technology without damaging the scientific infrastructure that has made biological research so vital to the health of the nation.

THE LIFE SCIENCES TODAY

The biological sciences have experienced enormous growth over the last century, fueled by a stream of discoveries—such as the principles of genetics, the structure of DNA, and the discovery of gene-splicing technologies. These have opened new fields of inquiry and provided the basis for myriad applications in industry, agriculture, and medicine. Among the technological breakthroughs in the life sciences, genetic engineering plays a particularly significant role.

Genetic engineering is a technique that permits the artificial modification and transfer of genetic material from one organism to another and from one species to another. This technology is used throughout the world to alter the protein produced by a gene and to design organisms with desirable traits for applications ranging from basic research and development activities to pharmaceutical and industrial uses. During the last 30 years, these recombinant techniques have spawned a vibrant biotechnology industry focused largely on the development of new pharmaceuticals to fight disease.[6] By 2000 the annual investment in the biotechnology industry peaked at nearly $29 billion, while employment in the biotechnology industry reached 191,000 by 2001.[7]

In response to the opportunities presented by these developments, the resources devoted to the life sciences have increased dramatically, making further discoveries possible. The government has funded biological research generously through the National Institutes of Health and National Science Foundation budgets, with few strings attached; private foundations and the pharmaceutical industry have also made major contributions. The number of PhDs awarded each year in the biological and agricultural sciences has increased steadily; 6,526 were awarded in 2001.[8]

This ever-expanding research activity has resulted in numerous new biopharmaceutical products that are transforming medicine. Examples include human recombinant insulin for the treatment of diabetes, a vaccine against hepatitis B, and medicines for diabetes, cancer therapy, arthritis, multiple sclerosis, cystic fibrosis, heart attacks, hemophilia, and sepsis. As knowledge of the human genome increases, it may even become possible to tailor pharmaceutical products not only to specific diseases but also to specific individuals. Throughout this process, the time between new discoveries and their applications has grown ever shorter. One example is the very short time it took the scientific community to identify the coronavirus as the causal agent of the newly emerging human disease, severe acute respiratory syndrome (SARS).

Biotechnology research is now a truly global enterprise. While industrialized countries such as the United States, the United Kingdom, Germany, Israel, and Japan may be the first to develop advanced research and technologies, other countries have a skill base that will enable broad domestic utilization of biological technologies.[9] For example:

> China has an aggressive program in plant biotechnology, and as of 2002 plans to increase funding by 400 percent by 2005. This energetic investment also exists in the Chinese private sector, and the national scientific establishment is attempting to lure foreign-trained scientists to return with lucrative financial packages. India is in the process of tripling funding to its national biotech center, and is promoting the development and use of genetically modified crops throughout Asia. Singapore has for many years made a practice of recruiting foreign scientists. Taiwan is investing large amounts in biotechnology and is seeking citizens to return home to build up biotechnology in academia and industry. A Brazilian coalition recently demonstrated sophisticated domestic use of biological technologies by successfully sequencing the plant pathogen *X[ylella] fastidiosa* in 2000.[10]

In addition to the dispersed research enterprise, publications and personnel are also widely spread. Well over 10,000 journals in the life sciences are published worldwide. Biological Abstracts, an international database on biology, clinical and experimental medicine, biochemistry, and biotechnology, provides coverage of over 6,000 active international journals and 14,000 archival titles from over 100 countries; Medline, the online service of the National Institutes of Health, provides abstract information for more than 4,600 biomedical journals published in the United States and 70 other countries; and PubMed currently provides full-text web access to 4,058 journals. According to Medline, the total number of scientific articles published in the peer-reviewed biomedical literature increased from 449,109 in 1998 to 491,620 in 2001. Given the global nature of the biotechnology research and development enterprise, it is unrealistic to think that biological technologies and the knowledge base upon which they rest can somehow be isolated within the borders of a few countries.

The rapid advance of scientific knowledge and applications owes much to a research culture in which knowledge and biological materials are shared among scientists and people move freely between universities, government agencies, and private industry. Large numbers of foreign graduate students and postdoctoral associates have been an essential ingredient in the success of the biological research enterprise. The scientific workforce is increasingly international; at the National Institutes of Health, for example, approximately 50 percent of the technical staff are non-U.S. citizens. Research results have been widely disseminated, so that even high school students now routinely perform experiments involving recombinant DNA techniques. In short, a dynamic national and international research enterprise has evolved, with an extraordinary record of achievement at multiple centers of excellence. These are values that should be preserved in any sensible policy for minimizing the risks associated with the misapplication of the fruits of the biotechnology enterprise.

THE DUAL USE DILEMMA

The regulation of dual use biotechnology research is a highly contentious technical, political, and societal issue. In the language of arms control and disarmament, dual use refers to technologies intended for civilian application that can also be used for military purposes. Technology involves more than just products; it also encompasses a means to produce and use products in such a way as to solve a problem. Thus, technology comprises "the ability to recognize technical problems, the ability to develop new concepts and tangible solutions to technical problems, the concepts and tangibles developed to solve technical problems, and the ability to exploit the concepts and tangibles in an effective way."[11]

The "general purpose clause" of the BWC prohibits the development, production, and stockpiling of biological weapons, but permits States that are parties to the treaty to conduct research activities for peaceful purposes or in order to defend or protect against BW agents.[12] Useful distinctions between permitted and prohibited activities at the level of basic research are difficult to make because biotechnology presents a classic example of the dual use dilemma. In the life sciences, for example, the same techniques used to gain insight and understanding regarding the fundamental life processes for the benefit of human health and welfare may also be used to create a new generation of BW agents by hostile governments and individuals. For the scientists and technicians involved in cutting-edge research and development in biology, biotechnology, medicine, and agriculture, this duality creates both uncertainties and ethical dilemmas. The duality between the purposes permitted and prohibited under the BWC is at the heart of this Committee's activities.[13]

Current research programs in universities, government laboratories, and pharmaceutical companies include experiments directed toward such goals as discovering vaccines for major diseases such as influenza, AIDS, and cancer; new antibiotics for both bacterial and fungal diseases; new sources of genes to protect crops against pests and diseases; and treatments for diabetes, stroke, and Alzheimer's disease. These research activities also include an intense effort to discover vaccines, antibiotics, and detection systems that would provide the defense against each of the select agents. But many of the same methods for developing attenuated live vaccines against viral diseases can have offensive applications as well.[14] The key issue is whether the risks associated with misuse can be reduced while still enabling critical research to go forward.

A BRIEF HISTORY OF MODERN BIOLOGICAL WARFARE

Of thousands of species of potentially pathogenic microorganisms, very few have been developed and deployed as biological weapons. As a society, we tend to think that biological and chemical warfare are recent threats to individuals and populations, but in reality, the offensive use of chemical and biological agents has its origins in antiquity (see Annex to this chapter). It has only been within the last century, however, that infectious disease agents have been seriously considered, on a continuing basis, as tools of war. Based on scientific discoveries during the late nineteenth and early twentieth centuries, biologists were able for the first time to identify, isolate, and culture disease-causing microbes under controlled conditions and use them to intentionally induce disease in a "naïve" host. "The foundations of microbiology pioneered by Louis Pasteur and Robert Koch offered new prospects for those interested in biological weapons because it allowed agents to be chosen and designed on a rational basis."[15]

Germany was accused of using disease-causing germs during World War I by infecting horses and mules with glanders—a highly infectious animal disease—and cattle with anthrax. German spies were caught in 1917 allegedly trying to spread anthrax bacteria among reindeer herds in the far northern portion of Norway, near the border with Russia.[16] These charges were confirmed when anthrax-laced sugar cubes—obtained from a Swedish-German-Finnish aristocrat arrested as a German agent in 1917—were found to be still viable after being stored in the archives of a Norwegian museum for the last 80 years.[17]

Over the past 60 years pathogens have been identified and perfected as strategic and tactical weapons. Every major combatant during World War II—including the United States, Great Britain, Canada, France, the former Soviet Union, Germany, and Japan—had some type of biological weapons program.[18] During the Sino-Japanese War (1937-1945), Japan repeatedly attacked China with the plague-causing bacteria *Yersinia pestis*, targeting some eleven cities. At least 700 Chinese reportedly died from plague alone,[19] although the number of Chinese civilians killed between 1933 and 1945 by Japanese germ warfare may be much higher.[20]

Japan's secret biological warfare program, Unit 731,[21] officially referred to as the Army Anti-Epidemic Prevention and Water Supply Unit, was located in a remote, high-security area in Japanese-occupied Manchuria, first in Harbin and then in Ping Fan. The Japanese perfected culture and dispersal techniques for a large number of biological agents. After the war the Japanese commander of Unit 731, General Shiro Ishii, traded research data, at the suggestion of his debriefers with the American occupation government in Japan, in exchange for a grant of immunity from war crimes prosecution. Information obtained from General Ishii later found its way to Camp Detrick, and is still held in the National Archives in the United States.[22]

The United States' offensive biological weapons program also had its origins in World War II. Begun in 1942 within the Chemical Warfare Service at Camp Detrick in Frederick, Maryland, the program's primary mission during World War II was biological warfare research on the causative agents of anthrax and botulism.[23] The main element for carrying out this program, the Special Projects Division of the Army Chemical Warfare Service, had at its peak 3,900 personnel, of which 2,800 were Army, nearly 1,000 Navy, and the remaining 100 civilian. The work was carried out at four installations. Camp Detrick was the parent research and pilot plant center. Field testing facilities were established in 1943 and 1944 in Mississippi and Utah, respectively, and a production plant was constructed in Indiana in 1944. All work, which was coordinated with Great Britain and Canada, was conducted under strictest secrecy.[24]

From the end of World War II until the U.S. decision to renounce its biological weapons program in 1969, this program developed and perfected offensive weapons capabilities for the Air Force, Navy, and the Central Intelligence Agency (CIA), utilizing a variety of human, animal, and plant pathogens.[25] "Between 1941 [sic] and 1969, the policy of the United States regarding biological warfare was first (to) deter its use against the United States and its forces, and secondly to retaliate if deterrence failed."[26]

The largest biological weapons complex ever created was in the former Soviet Union. Two main groups of facilities were involved in the research and development, production, and testing of biological weapons: a military-controlled system, which started in the 1920s, and Biopreparat, a top-secret program operating under civilian cover from 1972 until at least 1992,[27] despite the fact that the Soviet Union was an original signatory to and repository for the Biological Weapons Convention. As a result, the Soviet program not only caught up with but surpassed the U.S. program to become the most sophisticated biological weapons program in the world. Its size and scope were enormous; by the early 1990s more than 60,000 people were involved in the research, development, and production of biological weapons as well as the stockpiling of hundreds of tons of anthrax spores and tens of tons of other pathogens, including smallpox and plague.[28] In addition, it is now known that other state programs were involved in aspects of this effort including those of the Ministry of Health, Ministry of Agriculture, Ministry of Defense, KGB, and the Soviet Academy of Sciences.

U.S. POLICY AND THE CREATION OF THE BIOLOGICAL AND TOXIN WEAPONS CONVENTION

After intensive debate in the United Nations and domestic interagency review, President Richard Nixon on November 25, 1969 renounced the first use of lethal and incapacitating chemicals and stated that he would seek ratification of the Geneva Protocol by the U.S. Senate. (The Geneva Protocol of 1925 prohibits the use of chemical or biological materials in war, although it does not proscribe their acquisition or possession.) President Nixon also renounced the use of lethal bacteriological (biological) agents and weapons as well as all other methods of biological warfare, and directed the Defense Department to make recommendations for the disposal of existing BW stockpiles. He further stated that the United States would confine its biological agent and toxin research to defensive measures, such as immunization and safety. On February 14, 1970, this policy was extended to biological toxins regardless of their means of production.[29]

The United States decided to abandon its offensive biological weapons program, destroy its existing stockpiles of biological and toxin weapons, and convert the production facilities to other purposes because it was recognized that:

- Biological weapons could be as great a threat to large populations as nuclear weapons and that no reliable defense is likely;
- Biological weapons could be much simpler and less expensive than nuclear weapons to develop and produce; proliferation of biological weapons would therefore greatly increase the number of nations to which the populations of the United States and its allies [could] be held hostage;
- Our biological weapons program was pioneering an easily duplicated technology and was likely to inspire others to follow suit.[30]

The United States concluded that its biological weapons program was a substantial threat to its *own* national security and that one of the best ways to reduce this threat was not only to renounce biological weapons in this country but also to strengthen the international barriers to their proliferation.[31] The United States, the United Kingdom, and the former Soviet Union together were responsible for the effort to sponsor the Biological and Toxin Weapons Convention (BWC) of 1972—the first arms control agreement to ban outright an entire class of weapons.[32] The U.S. Senate ratified the BWC in 1975. To date, 162 countries have signed and 148 countries have ratified the BWC.

THE NEW THREAT

The revolution in biotechnology was just beginning when the BWC went into force in 1975. With the advent of the biotechnology revolution and the apparent proliferation of countries desiring to have a biological weapons capability, its signatories must reexamine the efficacy of the Convention in governing the use of disease as a method to spread terror.

The acquisition of biotechnology and biological weapons capability is considerably easier than was the case in the 1940s and 1950s. The explosion in biotechnologies and genetic engineering technologies—all of which have legitimate civilian applications—could empower a hostile agent. Gordon Oehler, director of the Non-Proliferation Center at the Central Intelligence Agency, testified before the Senate Armed Services Committee on March 27, 1996, and stated that there was "a continuing pursuit by many countries to acquire chemical and biological weapons and that [t]he chilling reality is that these materials and technologies are more accessible now than at any other time in history."[33]

The information to conduct genetic engineering research is easily accessible on the Internet. Moreover, the equipment and expertise to use this information to create novel agents are available globally. The international diffusion of knowledge and capabilities in biotechnology means that the capacity to carry out beneficial as well as harmful research activities is widely accessible, both to nations and to terrorist groups.

In this situation it is futile to imagine that access to dangerous pathogens and destructive biotechnologies can be physically restricted, as is the case for nuclear weapons and fissionable materials.[34] The nature of the biotechnology problem—indeed the nature of the biological research enterprise—is vastly different from that of theoretical and applied nuclear physics in the late 1930s. The contrast between what is a legitimate, perhaps compelling subject for research and what might justifiably be prohibited or tightly controlled cannot be made a priori, stated in categorical terms, nor confirmed by remote observation.

Matthew Meselson, a leading molecular biologist, gave a stark warning of the potential dangers posed by the destructive applications of biotechnology in May 2000:

> Every major technology—metallurgy, explosives, internal combustion, aviation, electronics, nuclear energy—has been intensively exploited, not only for peaceful purposes but also for hostile ones. Must this also happen with biotechnology, certain to be a dominant technology of the coming century? During the century just begun, as our ability to modify fundamental life processes continues its rapid advance, we will be able not only to devise additional ways to destroy life but ... also ... to manipulate it—including the processes of cognition, development, reproduction, and inheritance. A world in which these capabilities are widely employed for hostile purposes would be a world in which the very nature of conflict has radically changed. Therein could lie unprecedented opportunities for violence, coercion, repression, or subjugation.[35]

These dangers cannot be eliminated entirely since the fundamental knowledge from which they emerge is available around the world and the potential benefits of biotechnology for health promotion and national defense are too great to contemplate efforts to prohibit or reverse such research. But the potential adverse effects associated with the malicious exploitation of these technological advances cannot be ignored. Because of widespread moral repugnance against the production and use of chemical and biological weapons (CBW), the involvement of scientists and engineers in CBW research, development, and production is widely condemned.[36] History demonstrates, however, that without any military application in mind, research in biology may still contribute to the production of biological weapons.[37] As discussed earlier, the discovery and elaboration of the "germ theory of disease" in the nineteenth century led not only to better sanitation and hygiene practices but also to the intentional development of disease as a weapon in the twentieth century.

In discussing modifications of microorganisms that might have significance for bioweapons, Nixdorff and Bender[38] identified four classes of microbial manipulations that have been the subject of intense debate within and outside the scientific community:

1. The transfer of antibiotic resistance to microorganisms,
2. Modification of the antigenic properties of microorganisms,
3. Modification of the stability of microorganisms to the environment, and
4. The transfer of pathogenic properties to microorganisms.

Regarding these manipulations, they observed that:

All four kinds of manipulations are possible and are being carried out daily in research laboratories. Some of the most intensive research concerns the elucidation of the mechanisms of pathogenesis. This work is essential for combating infectious diseases. It is hoped that the production of more effective vaccines with [fewer] side effects, better diagnostics and new therapeutic drugs will result from this research. At the same time, it is feared that the advances in biotechnology can be misused to develop and produce biological weapons.[39]

The National Institutes of Health's (NIH's) recently released research priorities for countering bioterrorism identified several categories of research activities in immunology and genomics that would be considered "provocative" if conducted by a hostile or rogue government. These include efforts to "identify pathogen-induced immunoregulatory and immunosuppressive effects" as well as to "analyze gene expression of agents of bioterrorism."[40] John Gannon, former chairman of the National Intelligence Council and a former deputy director for intelligence at the CIA, observed that "the continuing revolution in science and technology will accentuate the dual use problem related to biotech breakthroughs in biomedical engineering, genomic profiling, genetic modification, and drug development.... Responsible scientists will have an extraordinary opportunity to improve the quality of human life across the planet. At the same time, terrorists and other evildoers may develop a powerful capability to destroy that life."[41]

RECENT EXAMPLES OF "CONTENTIOUS RESEARCH" IN THE LIFE SCIENCES

Biological weapons differ from other weapons systems in a number of important respects. They generally are based on naturally occurring pathogens that have coevolved along with their hosts to possess features such as high infectivity, ease of transmission, and virulence. As a corollary, how-ever, the effects of naturally occurring pathogens are limited by the evolu-tionary advantage gained by not eliminating their hosts. Among the many implications of the anticipated progress in biotechnology is the presump-tion that it may be feasible to create novel biological agents that are far more predictable and dangerous than any of the naturally occurring pathogens that have been developed as biological weapons in the past.[42] It may be difficult to engineer a more successful pathogen than those already present in nature that have been perfected by evolution for their niche in life. How-ever, application of the new genetic technologies makes the creation of "de-signer diseases" and pathogens with increased military utility more likely.[43]

There have been several recent examples of what Gerald Epstein of the Defense Threat Reduction Agency refers to as "contentious re-search"[44]—experiments that resulted in the creation of organisms or knowledge with "dual use" potential. The Australian *ectromelia* virus (mousepox) experiment; total synthesis of the poliovirus genome and re-covery of infectious virus, and the comparison of the immune response to a host defense function from *vaccinia* and smallpox have all attracted the attention of the scientific community, the media, the defense community, and policy analysts. Each is elaborated below.

The Mousepox Virus: A Case Study in Preconsideration

The mousepox virus: a case study in preconsideration. Probably the most celebrated recent case involving the dissemination of research with the potential for bioterrorist uses was the report of an unexpected effect of the bioengineering of a strain of *ectromelia* virus (mousepox) that was in-tended to help eradicate mice in Australia. The authors of the paper[45] had originally set out to make an infectious immunocontraceptive for wild mice by incorporating an ovary specific antigen, the mouse zona pellu-cida 3 (ZP3) glycoproteina gene into the genome of *ectromelia* virus. The authors subsequently sought to alter the ectromelia by adding an immunomodulator with the hope that this would increase the immune response of the infected mice to their fertilized eggs and thus make them permanently infertile. They drew upon previous published work by oth-ers with recombinant *vaccinia* virus in mice in which it had been shown that incorporating the gene for the immunomodulatory cytokine IL-4 into the viral genome and thus overexpressing it *in vivo* enhanced the viru-lence of *vaccinia* virus in mice. The increased virulence is probably due to suppression of the antiviral immune response mediated through compet-ing cytokines like IL-2, IL-12, and interferon gamma, which work by stimulating immune effector cells to kill virus-infected cells and thus con-trol the virus infection.

The authors of this study used standard and quite simple procedures for incorporating the IL-4 gene into the mousepox genome. They then demonstrated that this engineered mousepox virus was more virulent than the parent virus and killed 60 percent of infected mice, even if the mice were from a genetically resistant strain. Even more unexpected was their observation that mice that had been vaccinated and were completely resistant to the parent virus, and even to a more virulent strain of mousepox, were now killed by the IL-4 gene-expressing virus.

Some have felt that the publication of this paper provides a blueprint or road map for terrorists to engineer a more virulent strain of smallpox that could overwhelm the human immune system in even well-vaccinated individuals. The methods section of the paper illustrates how easy it is to make an IL-4 expressing orthopox virus. It has been suggested that either the paper should not have been published, or at the very least the "materials and methods" section of the manuscript should have been altered or omitted entirely from the published article. The authors were sensitive to these issues and consulted with their peers in the Pest Animal Control Cooperative Research Center at the Australian National University in Canberra about whether the paper should be submitted for publication. The manuscript was submitted in July 2000 to the *Journal of Virology*. The reviewers and editors expressed no concerns about potential misuse of any information in the manuscript and the article was published in February 2001. A retrospective review by the editor-in-chief following concerns raised after the article was published concluded that the journal was correct in its decision to publish.

This example illustrates the difficulty of attempting to censor either the initiation of research or the publication of results. The initial goals of the mousepox research were directed to control the population densities of a rodent pest in Australia. The studies were done on mice, and the virus itself, while related to smallpox, is not of any danger to humans. Thus, these studies had desirable scientific and societal goals and there was no obvious reason not to undertake them. Even in retrospect, the decision by informed scientists who had no vested interest in the work to approve publication seems appropriate. There were numerous examples in the published literature demonstrating the effects of cytokines like IL-4 on immune modulation. The authors of this study, therefore, were building upon an established literature in this field that is filled with similar findings on the effects of the decreased or increased levels of IL-4 and other immunomodulatory factors on the virulence of other viruses and many microorganisms. As previously noted, the design of the mousepox study built upon previously published studies in which *vaccinia* virus engineered to express IL-4 was studied in mice. There is also a relevant pre-existing literature in the field of oncology in which the increased expression of various cytokines incorporated into the fowlpox genomes and other orthopox viruses along with tumor antigens has been used to increase the immune response to tumors and decrease the immunogenicity of the viruses. The technique for incorporating new genes into the poxvirus genome had been published in many places. Thus there is little technical information that was not already abundantly available in the literature and well known to the scientific community.

The observation that even vaccinated mice were killed by the IL-4 expressing mousepox was a somewhat surprising finding that is of potential concern. However, since the ability of immunomodulatory factors to increase the virulence of this virus could have been predicted and the means to make such a virus were readily available, it was important to publicize that this strategy could overcome vaccination because it alerted the scientific community to such a possibility occurring either intentionally or spontaneously. First, knowledge of these experiments allows the scientific community to explore how to overcome such engineered viruses. It informs us of the fact that we should monitor cytokine levels in the blood of the initial cases of a highly virulent virus that is used in an attack. Second, it suggests that we should be prepared to treat infections caused by such an engineered virus with antibodies that inactivate the relevant cytokine, with gamma interferon that would counter the effect of IL-4, or with both. Finally, it is worth noting that this work was done outside the United States and could have been published in an Australian or European journal, illustrating the limits of national policies to address dual use concerns and, in this case specifically, the need to have international guidelines for the publication of manuscripts containing "sensitive" information.

Total Synthesis of the Poliovirus Genome and Recovery of Infectious Virus

Wimmer and colleagues[46] reported that they had reconstructed poliovirus from chemically synthesized oligonucleotides that were linked together and then transfected into cells. This report attracted considerable attention in the news media and concern in some segments of the public. The media treatment of the work suggested that this experiment proved that one could synthesize any virus from chemical reagents that can be purchased on the open market. This implication raised the public concern about bioterrorism because it suggested that the Wimmer experiment provided a recipe for terrorists to manufacture the virus. In response to the publication of this article in *Science*, in the 107th Congress, Representative Dave Weldon (R-FL) introduced H.Res. 514, which criticized the publication of this research because of its implications for compromising the national security interests of the nation. The Weldon resolution, which did not pass, went on to state the concern of the House of Representatives regarding the potential of the poliovirus paper to enable terrorists to synthetically create a human pathogen to release on the people of this country and further called upon the publishers and editors of scientific publications and the scientific community to establish ethical standards and exercise restraint in the dissemination of information of potential use to terrorists in the development of bioterrorism agents. The Weldon resolution also called upon the Executive Branch to "examine all policies, including national security directives, relevant to the classification or publication of federally-funded research to ensure that, although the free exchange of information is encouraged, information that could be useful in the development of chemical, biological, or nuclear weapons is not made accessible to terrorists or countries of proliferation concern."[47]

Many scientists concluded that the Wimmer experiment was neither a novel discovery nor a potential threat. The general principle that one could make live poliovirus from a DNA template was already known in 1981, when Baltimore and colleagues[48] reported that a DNA copy of the positive strand RNA genome of poliovirus could be taken up into living cells under appropriate conditions and result in the generation of encapsulated, infectious virus. These studies led to the ability to manipulate DNA copies of RNA viral genomes to generate preselected genetic changes. This technology bypasses the technical problems of working with RNA molecules and allows subsequent recovery of infectious virus. Subsequent research has succeeded in extending this technology to RNA viruses with larger positive strand genomes, negative polarity RNA, or segmented genomes.

Several points should be emphasized, however. Like the mousepox IL-4 experiment discussed above, the technology for producing and manipulating the genome of RNA viruses has been available in the literature for a long time. The ability to synthesize a poliovirus genome and recover infectious virus was regarded as a foregone conclusion. The Wimmer approach offers no technical advantage to a terrorist. And more importantly, in fact it is a very laborious and difficult way to accomplish this synthesis. The interesting scientific results from the Wimmer experiment were not its highly touted potential for bioterrorism, but rather the fact that the virus synthesized had significantly weakened pathogenicity as compared to wildtype strains of poliovirus. The decreased virulence is likely due to thirdbase and noncoding changes inserted as supposedly neutral markers.

Comparison of the Immune Response to a Virulence Gene from Vaccinia and Smallpox

Variola major virus causes smallpox, which has a 30-40 percent mortality rate, whereas *vaccinia* virus, which is used to vaccinate humans against smallpox, causes no disease in immunocompetent humans. In a paper that appeared in the *Proceedings of the National Academy of Sciences*, Rosengard and colleagues[49] investigated a possible basis for the difference in the putative virulence factor between the virus that causes human disease and the one used to vaccinate against the disease. Both viruses have an inhibitor of immune response enzymes—*vaccinia* virus complement control protein (VCP) and smallpox inhibitor of complement enzymes (SPICE). The authors focused on a comparison of the genes encoding this inhibitor. As live *variola* is not available for study, they used standard techniques to synthesize the *variola* SPICE gene. They found that *variola* SPICE has a greater degree of specificity for human complement and is nearly a hundredfold more active than VCP at inactivating this component of the human immune system (human complement component C3b). The authors suggested that the difference between VCP and SPICE could explain the difference in virulence between the two viruses and the restriction of *variola*'s host range to humans.

Some might argue that the Rosengard study is of greater concern than the previous two examples because it provides information on how to increase the virulence of *vaccinia* virus, and thus on how to convert a readily available agent that has minimal virulence into a virulent virus. A commentary written on this paper pointed out that it is very unlikely that *vaccinia* virus carrying SPICE in place of VCP would approach the pathogenicity of *variola*.[50] Furthermore, publication of the article alerted the community of scientists to this mechanism for virulence. This information should stimulate scientists in both the public and private sectors to identify compounds or immunization procedures that disable SPICE. These could form the basis for new treatments or vaccines both to immunize against the naturally occurring smallpox virus and to counteract the genetically engineered variety.

THE RESPONSE OF THE LIFE SCIENCES COMMUNITY TO PREVIOUS CHALLENGES

As the preceding examples make abundantly clear, there is an increasing awareness within and outside the scientific community of the dangers posed by the proliferation of biological weapons capabilities. This heightened awareness has also increased the collective concerns of this Committee and the scientific community about preventing the destructive applications of biotechnology research. This is not a completely new issue. When gene splicing technology was first reported, the scientific community at the time raised concerns that the technology might deliberately or inadvertently be used to create organisms with increased virulence or novel characteristics.[51] These possibilities eventually led to the 1975 Asilomar Conference, where scientists gathered to discuss the safety of manipulating DNA from different species.[52] The meeting resulted in the issuance by NIH of Guidelines for Research Involving rDNA Molecules (hereafter called the NIH Guidelines) in 1976 that regulated the conduct of NIH-sponsored recombinant DNA research and established a mechanism for reviewing proposed experiments in this field.

Just as the life sciences community with the Asilomar Conference stepped up to the challenge of responding to concerns that biology could set back rather than advance human welfare, so too the Human Genome Project created the ethical, legal, and social implications program to explore how advances in genetics intended to improve human health could proceed without undermining other dimensions of human well-being. National commissions and Congress continue to debate whether certain advances in biology should be pursued and published.[53]

The initial fears about the inadvertent creation of virulent microbes by gene splicing techniques have abated because of overwhelming scientific evidence to the contrary. There have been no reported cases of disease caused by recombinant microorganisms despite the widespread use of gene splicing techniques in academic laboratories and in the production of pharmaceuticals. In view of this experience, and the prospects for understanding the etiology of complex diseases and finding cures for them, the NIH has revised its Guidelines several times, with the net result being the elimination of the earlier prohibitions and the exemption from the Guidelines of essentially all recombinant DNA experiments except those that involve the molecular manipulation of human and restricted animal and plant pathogens.

COMMITTEE CHARGE AND PROCESS

Current policy at both the national and international levels may not be adequate to cope with the dangers inherent in the use and applications of genetic engineering. As discussed in greater detail in the following chapters, the United States has enacted legislation to provide for the physical security of select agents and screening of personnel. The Committee's proposed system for reviewing research projects and publications would complement and strengthen this statutory regime.

Internationally, however, protection against misuse of biotechnology is very uneven. The Biological and Toxin Weapons Convention, the centerpiece of biological weapons arms control, lacks effective verification and compliance measures. Moreover, it addresses only the actions of states and was never intended to guard against the development of a BW capability by individuals or nonstate actors (although national implementing legislation required by Article IV of the Convention could constrain the actions of individuals and groups within the state). In November 2001 the draft text for an international protocol covering compliance and verification measures was rejected by the United States. New, informal measures to strengthen the BWC being explored by expert groups and States parties are scheduled to continue over a period of three years (the first meetings were held in Geneva in August 2003). The measures include enactment of national criminal legislation supplemented by an enhanced extradition regime; security standards for pathogenic organisms; genetic engineering oversight; and international adoption of professional codes of conduct.[54] The hope is that these discussions will translate into coordinated action by the States parties, but at present only a few states have instituted security measures to protect against diversion and misuse of biotechnology.

The most elaborate treaty-based inspection procedures could not achieve effective restrictions at the level of basic research without severely restricting research in general. The inevitable diffusion of knowledge and capabilities has already demonstrated that the capacity to do harm is becoming globally available, both to state and nonstate actors. At the same time, developments in biotechnology are also capable of yielding great benefits, such as new treatments for many diseases. The distinction between the great opportunities and great dangers of biotechnology turns on assessing whether the risk(s) associated with the benefits of fundamental research outweigh the potential for misuse. The challenge to the scientific community, therefore, is to develop formal and informal processes and procedures to mitigate or minimize the destructive applications of advanced biotechnology without unduly restricting legitimate biotechnology research activities.

Beginning the process of addressing these challenges is the purpose of this study. Specifically, the Committee was charged to:

1. Review the current rules, regulations, and institutional arrangements and processes in the United States that provide oversight of research on pathogens and potentially dangerous biotechnology research, within government laboratories, universities and other research institutions, and industry. The review would focus on how choices are made about which research is and is not appropriate, and how information about relevant ongoing research is collected and shared.

2. Use the review to assess the adequacy of current U.S. rules, regula-

tions, and institutional arrangements and processes to prevent the destructive application of biotechnology research.

3. Recommend changes in those practices that could improve U.S. capacity to prevent the destructive application of biotechnology research while still enabling legitimate research to be conducted.

This report is part of a larger body of work that The National Academies have undertaken in recent decades on science and security issues, beginning with *Scientific Communication and National Security* in 1982 and continuing into the 1990s with the publication of *Chemical and Biological Terrorism: Research and Development to Improve Civilian Medical Response* (1999) and *Firepower in the Lab: Automation in the Fight Against Infectious Diseases and Bioterrorism* (2001). In response to the events of September 11th, the Academies undertook a comprehensive survey of the contributions that science and technology could make to countering terrorism; *Making the Nation Safer: The Role of Science and Technology in Countering Terrorism* was published in 2002. The report of its panel on bioterrorism, *Countering Bioterrorism: The Role of Science and Technology*, was published separately. In addition, the Institute of Medicine's Forum on Emerging Infections convened a 3–day workshop on *Biological Threats and Terrorism: Assessing the Science and Response Capabilities*, which was released as a workshop summary late in 2002. The 2002 report on *Countering Agricultural Bioterrorism*, a study already in progress prior to September 11th, enabled the Committee to focus its primary efforts on threats to human health. In the area of potential controls on information and data, the report on *Sharing Publication–Related Data and Materials: Responsibility of Authorship in the Life Sciences* (2003) is particularly relevant to the continuing concerns for ensuring the wide availability of the results of scientific research.[55] Information about current projects may be found on the Academies website http://www.nas.edu.

Committee Process

In creating the Ad hoc Committee on Research Standards and Practices to Prevent the Destructive Application of Biotechnology, the National Research Council (the operating arm of The National Academies) selected committee members representing a broad spectrum of backgrounds, expertise, and interests. Areas of expertise included molecular and cellular biology, virology, medicine, laboratory safety, international and regulatory law, bioethics, and defense policy (see Appendix B for biographical information on the members of the Committee). In addition, the Committee relied on the expertise and advice of representatives from the Executive Office of the President, governmental and nongovernmental technical and policy experts, as well as educators and private consultants. Information available from the open literature and materials submitted by experts were reviewed and considered during the Committee's deliberations (see Appendix C).

Even though the Statement of Task did not require the Committee to consider information control regimes for dissemination of information in the life sciences that could be exploited for nefarious purposes, the Committee concluded that this issue was implicit in the larger task before it and needed to be considered along with the regulatory environment for biotechnology research. An additional impetus for the Committee's consideration of information control regimes for unclassified research in the life sciences was the announcement by the White House shortly before the Committee's first meeting of its renewed interest in the application of "sensitive but unclassified information" control regimes for managing the dissemination of unclassified research that is financed by the federal government.[56]

Report Road Map

Chapter 2 reviews the current domestic and international rules, regulations, and institutional arrangements and processes that provide oversight of research on pathogens and potentially dangerous biotechnology research within government laboratories, universities and other research institutions, and industry. Chapter 3 reviews the existing and emerging regulatory environment governing the control of information related to biological research. Chapter 4 presents the Committee's conclusions and recommendations about the ways in which the current regulatory environment for genetic engineering research might be enhanced while allowing the scientific enterprise to continue its essential activities.

ANNEX:
BIOLOGICAL WARFARE IN HISTORY

People figured out how to intentionally spread illnesses long before naturalists came up with the discovery that germs cause disease. Among the older military techniques that can be claimed as biological warfare is the use of corpses of humans or animals to befoul wells or other sources of drinking water.[57] While the principal objective was thought to be the denial of clean water to the enemy, a secondary effect was to spread disease among people and animals that consumed the contaminated water.[58] One of the earliest recorded instances of biological warfare occurred in 600 BC, when the Athenian leader Solon poisoned the water supply in the city of Kirrha with the noxious roots of the *Helleborus* plant—a primitive but effective biological toxin of plant origin. The Greeks and Romans may have used human and animal corpses to poison drinking water wells.

Alexander the Great is thought to have catapulted the bodies of dead men over the walls of besieged cities, possibly as a means of spreading disease and inciting terror among their inhabitants.[59]

A related technique, used in the Middle Ages, was to deliberately leave dead human or animal corpses behind in areas that would be occupied shortly by invading troops; catapults were used as well.[60] In 1346, invading Tartars intent on controlling the Silk Road trade attacked the Black Sea port of Caffa—at the time occupied by the Genoese. The Tartar army, already exposed to the Black Death, hurled plague-infested cadavers over the impregnable walls of Caffa to infect the enemy population.[61] It is usually reported[62] that the fleeing Genoese brought the Black Death with them—via plague-infested rodents, along shipping routes to Sicily, Sardinia, Corsica, and Genoa—and from there it spread overland throughout Italy and Europe. It is considered equally likely, however, that the entry of plague into Europe from the Crimea occurred independent of this event.[63] Over a four-year period, the plague eventually caused 25 million deaths—one-third of Europe's population at the time. Population losses were probably much higher in the French Mediterranean coastlands and in northern Italy.[64]

During the seventeenth and eighteenth centuries, French and British soldiers and civilians are alleged to have deliberately infected North American Indian populations with European diseases. "(T)he use of smallpox as a weapon may have been widely entertained by British military commanders and may have been employed without scruple when opportunity offered, possibly on a number of occasions."[65] During the French and Indian Wars, for example, Sir Jeffrey Amherst, commander-in-chief of the British forces, was concerned that his troops west of the Allegheny Mountains were in danger of being overrun by Indians. He wrote to the commander of the garrison at Fort Pitt on the Pennsylvania frontier and urged that smallpox be spread among the disaffected tribes.[66] In June 1763, Captain Ecuyer of the Royal Americans met with two Indian chiefs under a pretense of friendship and gave them blankets that had been taken from a smallpox hospital. During the following months, according to historians of the episode, many Indians suffered and died as "smallpox raged among the tribes of the Ohio."[67] During the 1800s, U.S. government agents were alleged to have deliberately infected the Plains Indians by giving them trading blankets infected with the deadly disease, decimating the population.[68]

NOTES

[1] Interview with President Boris Yeltsin, *Rossiskiye Vesti*, May 27, 1992. In Foreign Broadcast Information Service. Central Intelligence Agency. Washington, D.C.: FBIS-SOV-92-103.

[2] John Holum, then director of the Arms Control and Disarmament Agency, listed a dozen unspecified countries in 1996 as possessing or pursuing BW capabilities, commenting that this was twice as many as in 1975 when the BWC entered into force. Remarks to the Fourth Review Conference of the Biological Weapons Convention in Geneva, Switzerland, November 26, 1996. In May 2002 Under Secretary of State John Bolton named Iran, Iraq, North Korea, Libya, Syria, and Cuba as states the United States was certain possessed or were actively seeking BW. "Beyond the Axis of Evil: Additional Threats from Weapons of Mass Destruction," remarks to the Heritage Foundation, Washington, D.C.

[3] President George W. Bush. June 1, 2002. Remarks at the Graduation Exercise of the United States Military Academy. Available at http://www.whitehouse.gov/news/releases/2002/06/20020601-3.html.

[4] Intelligence estimates prior to the war concluded that Iraq had stocks of biological weapons. "We judge that Iraq has continued its weapons of mass destruction (WMD) programs in defiance of UN resolutions and restrictions. Baghdad has chemical and biological weapons as well as missiles with ranges in excess of UN restrictions; if left unchecked, it probably will have a nuclear weapon during this decade." National Intelligence Estimate, "Iraq's Continuing Programs for Weapons of Mass Destruction," October 2002. Available at http://www.ceip.org/files/projects/npp/pdf/Iraq/declassifiedintellreport.pdf. An "interim progress report" on the search for banned weapons of mass destruction in Iraq released on October 2, 2003 revealed no stockpiles of such weapons, though it did cite "rudimentary" traces of weapons programs, concealed equipment, and so forth. A copy of the unclassified statement, presented by David Kay and made available by the CIA, may be found at: http://www.fas.org/irp/cia/product/dkay100203.html.

[5] Zanders, J.P. 2002. Introduction in "Ethics and Reason in Chemical and Biological Weapons Research," *Minerva* (special Issue); 40:5.

[6] For purposes of this report, biotechnology is broadly defined to include "any technique that uses living organisms (or parts of organisms) to make or modify products, to improve plants or animals, or to develop microorganisms for specific use." Although the fields of biotechnology have expanded greatly in the last three decades, the term is more suitable for this study than biology or life science. Office of Technology Assessment (1988): *New Developments in Biotechnology: Field-Testing Engineered Organisms: Genetic and Ecological Issues*, May. NTIS Order #PB88-214101.

[7] See U.S. Biotech Employment Chart, available at http://www.bio.org/investor/signs/200210emp.asp.

[8] "Doctorates awarded by field of study and year of doctorate, 1999-2001," *Science and Engineering Doctorate Awards: 2001*. National Science Foundation, October 2002, p. 5.

[9] Carlson, R. 2003. The Pace and Proliferation of Biological Technologies, *Biosecurity and Bioterrorism: Biodefense Strategy, Practice and Science*. 1 (3):203-215.

[10] *Ibid.*

[11] Autio, E., and T. Laamanen. 1995. "Measurement and evaluation of technology transfer: Review of technology transfer mechanisms and indicators." *International Journal of Technology Management*. 10(7/8): 647 as cited in P. Zanders, op. cit., p. 6.

[12] United Nations. 1972. Convention on the Prohibition of the Development, Production and Stockpiling of Bacteriological (Biological) and Toxin Weapons and on Their Destruction. United Nations General Assembly Resolution 2826 (XXVI) (New York: United Nations). The BWC recognizes that the equipment and materials used to produce BW agents are almost entirely dual use, having legitimate commercial as well as military applications. For this reason, the treaty specifically prohibits only those activities involving pathogens that "cannot be justified for prophylactic, protective, and other peaceful purposes."

[13] Under the general purpose criterion, it is not the objects themselves but rather the purposes for which they may be applied that are prohibited. In this way, it is not necessary to ban dual use technologies that have legitimate purposes but that can also be applied to develop or produce BW. By using the general purpose criterion, the scope of the prohibition is comprehensive, because Art. 1 of the BWC lists the purposes that are not prohibited. Nixdorff, K., and W. Bender. 2002. "Ethics of University Research, Biotechnology and Potential Military Spin-Off," *Minerva* (special Issue):40, fn. 2, p. 15 and fn. 18, p. 19.

[14] Huxoll, D. 1989. "Biological weapons proliferation and the new genetics," Testimony before the Senate Committee on Governmental Affairs and its Permanent Subcommittee on Oversight and Investigations. May 17. Senate Hearing, 101-744; 101st Congress, 1st session.

[15] Frischknecht, F. 2003. "The history of biological warfare: Human experimentation, modern nightmares, and lone men in the twentieth century," *EMBO Reports* 4 (special issue): S47.

[16] Wheelis, M. 1999. "Biological sabotage in world war I," in Geissler E. and J.E. van Courtland Moon, eds., "Biological and Toxin Weapons: Research, Development and Use from the Middle Ages to 1945," SIPRI, 18 (London: Oxford University Press), p. 52.

[17] Redmond, C., et al. 1998. "Deadly relic of the great war," *Nature,* 393:747-748.

[18] Geissler, E., and J.E. van Courtland Moon, eds. 1999. "Biological and Toxin Weapons: Research, Development and Use from the Middle Ages to 1945," SIPRI, 18 (London: Oxford University Press).

[19] Williams, P., and D. Wallace. 1989. *Unit 731: The Japanese Army's Secret of Secrets.* (London: Hodder and Stoughton), pp. 280-281; and Harris, S.H. 1994. *Factories of Death: Japanese Biological Warfare, 1932-45, and the American Cover-Up* (London: Routledge).

[20] "Chinese Civilians Sue Over WWI-Era Japanese Biological Weapons Activities" CBW Chronicle III (3):December 2001. "http://www.stimson.org/cbw/?sn=cb20020112244" http://www.stimson.org/cbw/?sn=cb20020112244.

[21] There were at least four operational units of the Japanese secret biological warfare complex: Unit 731, located in Ping Fan, Unit 100 in Changchun, Unit 9420 in Singapore, and Unit Ei 1644 in Nanking. There is also some evidence that the Japanese had an epidemic prevention center—a euphemism for BW research on tropical diseases—in Rangoon, Burma. Each unit had 10–15 individual facilities located within and outside mainland China. See Williams, P. and D. Wallace, 1989, *Unit 731: The Japanese Army's Secret of Secrets.* (London: Hodder and Stoughton), p. 280-281; and Harris, S.H. 1994, *Factories of Death: Japanese Biological Warfare, 1932-45, and the American Cover-Up* (London: Routledge).

[22] *Ibid.*

[23] Bernstein, B. 1988. "America's biological warfare program in the Second World War," *Journal of Strategic Studies* 11 (September):292-317, especially p. 304 and 308-310. In addition to *Bacillus anthracis* and *Clostridium botulinum*, pathogens studied at Camp Detrick included the causative agents of: glanders; brucellosis; tularemia; melioidosis; plague; psittacosis; coccidiomycosis; a variety of plant pathogens including the causative agents for rice blast; rice brown spot disease; late blight of potato; and cereal stem rust. Animal and avian pathogens studied included rinderpest virus, Newcastle disease virus, and fowl plague virus. *The Problem of Chemical and Biological Warfare*, SIPRI, I (London: Oxford University Press), 1971, p. 122. See also Cochrane, R.C. 1947. "Biological Warfare Research in the United States" in *History of the Chemical Warfare Service in World War II* (1 July 1940 – 15 August 1945), Vol. II (declassified). Historical Section, Office of Chief, Chemical Corps.

[24] U.S. Department of the Army. 1977. *U.S. Army Activity in the U.S. Biological Warfare Programs* I; (unclassified) February 24, p. 1-3.

[25] *Ibid.*

[26] *Ibid.*, p. iii.

[27] Alibek, K., and S. Handelman. 1999. *Biohazard: The Chilling True Story of the Largest Covert Biological Weapons Program in the World – Told from the Inside by the Man Who Ran It* (New York: Random House).

[28] For personnel numbers, see Leitenberg, M. 1993, "The Conversion of Biological Warfare Research and Development Facilities to Peaceful Uses," in *Control of Dual-Threat Agents: The Vaccines for Peace Programme*, SIPRI Chemical and Biological Warfare Series, 15 (London: Oxford University Press). For the environmental impacts associated with biological weapons field testing see Choffnes, E. 2001, "Germs on the loose," *The Bulletin of the Atomic Scientists* 57 (March/April): 57-61.

[29] U.S. Department of the Army. 1977. *U.S. Army Activity in the U.S. Biological Warfare Programs*, Vol. 1; Feb. 24, p. 7-1.

[30] Meselson, M. 1989. Testimony to the U.S. Senate Committee on Governmental Affairs and its Permanent Subcommittee, Hearings on *Global Spread of Chemical and Biological Weapons."* May 17 (Washington, D.C.: U.S. Government Printing Office, 1990), pp. 498-511. See also National Security Council, 1969. "U.S. Policy on Chemical and Biological Warfare Agents," report submitted by the Interdepartmental Political-Military Group in response to NSSM 59, November 10. Available at http://www.gwu.edu/~nsarchiv/NSAEBB/NSAEBB58/#docs.

[31] *Ibid.*

[32] Unlike the Nuclear Non-Proliferation Treaty and the Chemical Weapons Convention, however, the BWC does not have formal mechanisms to monitor or enforce compliance. The BWC also has no international secretariat or inspectorate to oversee or verify its implementation. Thus, although the treaty enshrines a norm of international behavior, it lacks the capacity to enforce these prohibitions.

[33] Biotechnology and Genetic Engineering: Implications for the Development of New Warfare Agents–1996; Executive Summary available at http://www.acq.osd.mil/cp/biotech96/xsum.pdf.

[34] This is not recognized by some of those concerned about the proliferation of biological weapons or bioterrorism resulting from the diffusion of advanced biotechnology research. The Counterterrorism Act of 2000, for example, which passed

the Senate but not the House in the 106th Congress, cited the recommendation of the National Commission on Terrorism that "the standards for the storage, transport, and handling of biological pathogens should be as rigorous as the current standards for the physical protection of critical nuclear materials." Congress, Senate. *Counterterrorism Act of 2000*, 106th Congress, 2nd session, S. 3205.

[35] Meselson, M. 2000. "The Problem of Biological Weapons," remarks at the Symposium on Biological Weapons and Bioterrorism, National Academy of Sciences, May 2.

[36] Nixdorff, K., and W. Bender. 2002. "Biotechnology, Ethics of Research, and Potential Spin-off," *INESAP Information Bulletin*, 19 (March): p. 19-22.

[37] *Ibid.*

[38] *Ibid.*

[39] *Ibid.*

[40] Fauci, A. 2002. "Defining 'Sensitive' Information in the Life Sciences," oral presentation to the committee, September 9.

[41] Gannon, J.C. 2001. "Viewing Mass Destruction Through A Microscope," *New York Times*, Section E, p. 10, October 11.

[42] Developers of BW agents would strive for the greatest possible degree of predictability in infectiousness, virulence, and other militarily relevant characteristics.

[43] Speaking at the conference on the future of weaponry, Professor Kathryn Nixdorff, of the University of Darmstadt, said that dangerous microorganisms had already been produced inadvertently during attempts to modify vaccines and viruses. In the past 30 years biotechnology has been revolutionized by molecular biology and genetic engineering. These techniques, used to control infectious diseases, can also be used to create more effective biological weapons. See Hearst, D. 2003. "Smart bio-weapons are now possible," *The Guardian*. May 20. Available at http://www.guardian.co.uk/uk_news/story/0,3604,959473,00.html.

[44] Epstein, G.L. 2001. "Controlling biological warfare threats: Resolving potential tensions among the research community, industry, and the national security community," *Critical Reviews in Microbiology*, 27(4):321-354. Epstein defines "contentious research" as "fundamental biological or biomedical investigations that produce organisms or knowledge that could have immediate weapons implications, and that therefore raise questions concerning whether and how that research should be conducted and disseminated."

[45] Jackson, R.J., A.J. Ramsay, C.D. Christensen, S. Beaton, D.F. Hall, and I.A. Ramshaw. 2001. "Expression of mouse interleukin-4 by a recombinant ectromelia virus suppresses cytolytic lymphocyte responses and overcomes genetic resistance to mousepox," *Journal of Virology* 75:1205-1210.

[46] Cello, J., A.V. Paul, and E. Wimmer. 2002. "Chemical synthesis of poliovirus cDNA: Generation of infectious virus in the absence of natural template," *Science Online*, July 11. Available at http://www.sciencemag.org/cgi/ontent/full/297/5583/1016.

[47] Shea, D. 2003. "Balancing Scientific Publication and National Security Concerns: Issues for Congress." (Washington, D.C.: Congressional Research Service, Report Number RL31695), January 10.

[48] Racaniello, V.R., and D. Baltimore. 1981. "Cloned Poliovirus Complementary DNA Is Infectious in Mammalian Cells," *Science* 214:916-19.

[49] Rosengard, A.M., Y. Liu, Y.Z. Nie, and R. Jimenez. 2002. "Variola virus immune evasion design: Expression of a highly efficient inhibitor of human complement," *Proceedings of the National Academy of Sciences*, 99: 8808-8813.

[50] Lachmann, P.J. 2002. "Microbial subversion of the immune response," *Proceedings of the National Academy of Sciences* 99: 8461-8462.

[51] Wade, N. 1980. "Biological Weapons and Recombinant DNA," *Science* 208:271; S. Budianski. 1982. "US Looks to Biological Weapons. Military Takes New Interest in DNA Devices," *Nature* 297: 615-616.

[52] It should be noted that the Asilomar Conference addressed only the *accidental* creation of recombinant microorganisms with increased virulence and other dangerous properties. It did not address the *deliberate* creation of such organisms for offensive applications in warfare and terrorism.

[53] Kennedy, D. 2003. "Two Cultures" and "Statement on Scientific Publication and Security," *Science* 299 (5610):1148-1150. Available at http://www.sciencemag.org/content/vol299/issue5610/index.shtml.

[54] U.S. Department of State. 2001. "New Ways to Strengthen the International Regime Against Biological Weapons," Fact Sheet, Bureau of Arms Control, Washington, D.C., October 19. Available at http://www.state.gov/t/ac/bw/fs/2001/7909.html.

[55] All of these reports are published by The National Academies Press. Information about these and other reports may be found at http://www.nap.edu, which may be searched by subject matter and by report title.

[56] Card, A.H. Jr. 2002. "Action to Safeguard Information Regarding Weapons of Mass Destruction and Other Sensitive Documents Related to Homeland Security," March 19. Avaliable at http://www.fas.org/sgp/bush/wh031902.html.

[57] Stockholm International Peace Research Institute. 1971. "Instances and allegations of CBW, 1914 -1970," SIPRI, The Problem of Chemical and Biological Warfare, p. 214 in Vol. 1: *The Rise of CB Weapons* (Almqvist & Wiksell: Stockholm).

[58] *Ibid.*, p. 215.

[59] Glenn, J. 1989. "Biological weapons proliferation and the new genetics," Chairman's Opening Statement; Senate Committee on Governmental Affairs and its Permanent Subcommittee on Oversight and Investigations. May 17. Senate Hearing, 101-744; 101st Congress, 1st session.

[60] SIPRI, op. cit., p. 215.

[61] According to one account the plague in Caffa "might have spread naturally because of unhygienic conditions in the beleaguered city." Frischknecht, F. 2003. "The history of biological warfare: Human experimentation, modern nightmares, and lone madmen in the twentieth century," *EMBO Reports* 4 (special issue): S47.

[62] Wheelis, M. 2002. "Biological warfare at the 1346 siege of Caffa." *Emerging Infectious Diseases* 8(9):1971. Available at http://www.cdc.gov/ncidod/EID/vol8no9/pdf/01-0536.pdf. See also Wheelis, M. 1999. "Biological Warfare Before 1914," in E. Geissler and J.E. van Courtland Moon, eds, "Biological and Toxin Weapons: Research, Development and Use from the Middle Ages to 1945," SIPRI, 18 (London: Oxford University Press).

[63] Ibid.; see also, Frischknecht, F. 2003. "The history of biological warfare: Human

experimentation, modern nightmares, and lone madmen in the twentieth century," EMBO Reports. 4 (special issue): S47.

[64]Italian records are potentially very rich but have only begun to be carefully studied. Cf. Bowsky, W.M. 1964. "The impact of the Black Death upon Sienese government and society," *Speculum*, 39:1-34. D. Herlihy. 1966. "Population, plague and social change in rural Pistola, 1201-1430." *0*, 18:225-244. Some French towns also have abundant notarial records that can yield data on plague losses. Cf. Emery, R.W. 1967. "The black death of 1348 in Perpignan." *Speculum*, 42: 611-623. Emery estimated a die-off of 58-68 percent among the notaries of Perpignan from the plague. As cited in, McNeill, W.H. 1976. *Plagues and Peoples*, (Garden City, New York: Anchor Press). See also, Wheelis, M. 2002. "Biological warfare at the 1346 siege of caffa." *Emerging Infectious Diseases* 8(9):1971. Available at http://www.cdc.gov/ncidod/EID/vol8no9/pdf/01-0536.pdf. "The claim that biological warfare was used at Caffa is plausible and provides the best explanation of the entry of plague into the city. This theory is consistent with the technology of the times and with contemporary notions of disease causation; however, *the entry of plague into Europe from the Crimea likely occurred independent of this event.*" (emphasis added).

[65] Fenn, E. 2000. "Biological warfare in eighteenth century America: Beyond Jeffrey Amherst." *Journal of American History* 86:1552-1558, and Fenn, E.A. 2001. *Pox Americana: The Great Smallpox Epidemic of 1775-1782*. (New York: Hill & Wang Publishers).

[66] Op. cit., Fenn, E.A. 2001, and E.W. Stearn and A.E. Stearn. 1945. *The Effect of Smallpox on the Destiny of the Amerindian*. (Boston: Bruce Humphries Publishers), pp. 45-55.

[67] It is also possible that by the time of this intentional introduction of smallpox among the tribes of the Ohio that there was a concurrent outbreak of smallpox among these tribes. On the topic of smallpox blankets and Native Americans, see Wheelis, M. 1999. "Biological Warfare Before 1914," in Geissler, E. and J.E. van Courtland Moon, eds., "Biological and Toxin Weapons: Research, Development and Use from the Middle Ages to 1945," SIPRI, 18 (London: Oxford University Press). Also Fenn, 2000, op. cit.

[68] See Wheelis, M. 1999. "Biological Warfare Before 1914," in Geissler, E. and J.E. van Courtland Moon, eds., "Biological and Toxin Weapons: Research, Development and Use from the Middle Ages to 1945" SIPRI, 18 (London: Oxford University Press). Also Fenn, 2000, op. cit.

2
The Evolving Regulatory Environment for Life Sciences Research in the 21st Century

INTRODUCTION

The regulatory environment for the life sciences has been developed over the course of five decades. Responsibility for regulation in the United States of various aspects of biotechnology research in the life sciences is shared among a number of federal agencies, ranging from the Department of Agriculture (USDA) and the Environmental Protection Agency (EPA) to the Nuclear Regulatory Commission (NRC). The National Institutes of Health (NIH), for example, sets standards and procedures for the research it funds on recombinant DNA (rDNA). Research on human gene therapy is a special case, with both NIH and the Food and Drug Administration (FDA) conducting reviews prior to the initiation of research. While review is mandatory for NIH-funded research, industry often seeks review voluntarily. The Centers for Disease Control and Prevention (CDC) sets standards for the handling and transport of some especially dangerous biological pathogens. The NRC has responsibility for regulations to control the receipt, possession, use, transfer, and disposal of radioactive materials by research institutions. The USDA has broad responsibility for biotechnology research related to plants and animals. Industries involved in biotechnology research generally have internal procedures to review potential research protocols, although these vary considerably within and between industrial and commercial facilities. Important aspects of their work such as clinical or field trials are also subject to regulation by federal agencies, in particular the FDA and the EPA. Universities have various methods for reviewing and approving potentially contentious research. Professional societies address questions of ethics and norms for research.

Until the mid-1990s the regulatory environment focused on protecting the public health and general environment from biological hazards associated with possible exposures to human pathogens via interstate transport, recombinant organisms, and containment of recombinants and their products so that inadvertent or deliberate releases of these materials to the environment would be within acceptable limits. The regulatory environment for microbial hazards also encompasses the importation of nonnative plant and animal pathogens.

Following the historic Asilomar Conference in 1975, the NIH, in 1976 published the Guidelines for Research Involving rDNA Molecules (hereinafter referred to as the NIH Guidelines or the Guidelines). The NIH Guidelines described four levels of combinations of laboratory practices, containment equipment, and facility safeguards that were thought to be appropriate for the safe use and physical containment of rDNA molecules in research. The four levels, P1 to P4, provide increasing levels of physical protection against personnel contact with or accidental release to the environment of genetically engineered microorganisms.[1]

The CDC and the NIH encouraged the life sciences community to participate in a collaborative initiative to develop consensus guidelines to safeguard worker safety and public health from hazards associated with the possession and use of human pathogens in microbiological and biomedical laboratories. This initiative resulted in the publication by CDC and NIH in 1984 of Biosafety in Microbiological and Biomedical Laboratories (hereinafter referred to as the BMBL).[2] These consensus guidelines also established four ascending levels of physical containment using the terminology Biosafety Level 1-4. The combinations of standard and special microbiological practices, safety equipment, and facilities for each level are similar to those of the NIH Guidelines. Specific recommendations for appropriate practices, equipment, and facility safeguards are given in the BMBL for pathogens that meet one or more of three criteria: the pathogen is a proven hazard to laboratory personnel working with infectious materials; the potential for laboratory-acquired infection is high even in the absence of previously documented laboratory-associated infections; or the consequences of infection are grave. The recommendations are advisory and are intended to provide a voluntary guide or code of practice for investigators who possess and use human pathogens in their research activities.

The Recombinant DNA Advisory Committee (RAC) and the BMBL process have been highly successful. Laboratory-acquired infections from exposure to biological agents known to cause disease are infrequent. There are no reports that the possession and use of biological agents and toxins for research, education, and other legitimate purposes endangers the public health. The fourth edition of the BMBL states: "Experience has demonstrated the prudence of the Biosafety Level 1-4 practices, procedures, and facilities described for manipulations of etiologic agents and laboratories settings and animal facilities. Although no national reporting system exists for reporting laboratory-associated infections, anecdotal information suggests that strict adherence to these guidelines does contribute to a healthier and safe work environment for laboratorians, their coworkers, and the surrounding community."[3] This experience indicates that compliance with voluntary guidelines can achieve safety in research and clinical laboratories and protect the public health without significantly restricting the pursuit of science.

Spurred by rising concerns about bioterrorism, we are now witnessing a transition from an environment based upon voluntary compliance with recommended practices to a greater number of statutes and regulations, particularly for control of biological materials and personnel. It took the United States three years to ratify the 1972 Biological Weapons Convention (BWC) and 17 years for the United States Congress to pass legislation making the provisions of the BWC binding on all Americans.[4] Not much changed until 1996 when, with the passage of the Antiterrorism and Effective Death Penalty Act, new regulatory controls were enacted swiftly regarding transfers of dangerous pathogens.[5] Less than a year following the terrorist attacks on September 11, 2001 and subsequent anthrax mailings, two major pieces of legislation were passed by Congress and signed into law—"The Uniting and Strengthening America by Providing Appropriate Tools Required to Intercept and Obstruct Terrorism of October 2001"[6] (hereinafter, the PATRIOT Act), and "The Public Health Security and Bioterrorism Preparedness and Response Act" of June 2002[7] (hereinafter, the Bioterrorism Response Act).

The PATRIOT Act makes it illegal in the United States for anyone to possess any biological agent, including any genetically engineered organism created by using rDNA technology, for any inappropriate reason. The Act also prohibits the transfer or possession of a listed biological agent or toxin by a "restricted person."[8] A "restricted person" is not permitted to ship or transport via interstate or foreign commerce, or possess, or receive any biological agent or toxin that has been shipped or transported in interstate or foreign commerce, if the biological agent or toxin is listed as a select agent.[9]

The Bioterrorism Response Act added new requirements for the secretaries of the Departments of Agriculture and Health and Human Services to consider in listing agents and in preventing unlawful access to agents during transfers.[10] The statute also establishes new requirements for registration with the appropriate secretary concerning possession and use of listed agents and toxins, including "information regarding the characterization of listed agents and toxins to facilitate their identification, including their source; and safeguard and security requirements for registered persons." The law also requires the secretary to establish rules that provide appropriate physical security requirements for listed agents and for the Department of Justice—through the Federal Bureau of Investigation (FBI)—to conduct background investigations on individuals who are permitted access to select agents or who work in a facility where select agents are stored. The security provisions in the Bioterrorism Response Act are radically transforming the life sciences research environment in the United States from one that is basically open to one that excludes, based upon criteria stipulated in the PATRIOT Act, certain individuals from access to and research on certain listed agents. The FBI provisions, which went into force without public notice and comment rulemaking, prescribe the collection of pertinent background information on individuals; who may access, use, receive or transfer select agents, and the release and disclosure of that information to other entities as described in Section IV in the FBI Information Form (FD-961).[11] These provisions have raised concerns that qualified individuals may be discouraged from conducting biomedical and agricultural research of value to the United States because of the apparent infringement of these rules on individual liberties under the Fourth Amendment.

The next section of this report expands on the brief descriptions above to give a more complete picture of the current system of regulations and voluntary practices that govern research in biotechnology. It adds discussion of the growing web of controls over foreign nationals seeking to work, study, or participate in scientific activities in the United States and of the various codes of professional conduct that are a fundamental part of the self-governance of scientific practice. As noted at the beginning of this chapter, at present this system is focused on occupational safety and health and on environmental protection, but increasingly, additional efforts are being made to control access to biological materials that might be used by terrorists. With the exception of research involving human subjects, the system is not intended to provide oversight of research in the sense of making decisions about whether particular projects or experiments are appropriate.

THE U.S. REGULATORY ENVIRONMENT

Oversight of Genetic Engineering Research

Chapter 1 mentioned the response of the life sciences community in the mid-1970s to concerns about the potential unknown risks inherent in research involving the new field of genetic engineering. The Asilomar process led to the NIH assuming responsibility for promoting safe conduct of such experiments and to the subsequent publication of the NIH Guidelines. The Guidelines are designed to address the risks to public health and the environment associated with exposure to either rDNA molecules or organisms or viruses containing such materials.[12] The NIH Guidelines are applicable to all rDNA research within or outside the United States or its territories where the research is conducted at an institution that receives any support for the research from the NIH, including research performed directly by NIH.

Institutions that are recipients of NIH support for rDNA research must establish an Institutional Biosafety Committee (IBC) as part of their compliance with the NIH Guidelines. Further, as part of documenting that they have established a properly constituted IBC, institutions must register the IBC with the NIH Office of Biotechnology Activities (OBA).[13] IBCs are the cornerstone of institutional oversight of rDNA research.

An IBC is a review body appointed by an institution to review and approve potentially biohazardous lines of research relating primarily to rDNA research. IBCs were originally established to provide local, institutional oversight of nearly all forms of research utilizing rDNA. On behalf of the institution, IBCs review rDNA research projects for compliance with the NIH Guidelines. Over time, the role of the IBCs at many institutions has been expanded to include review and oversight of a variety of experimentation that involves biological materials (e.g., infectious agents) and other potentially hazardous agents (e.g., carcinogens).

While an IBC must consist of at least five members there is no upper limit on the number of members. Every IBC is required to have two members not affiliated with the institution who represent the interests of the surrounding community with respect to health and protection of the environment. These may be officials of state or local public health or environmental protection agencies, members of other local governmental bodies, or persons active in medical, occupational health, or environmental concerns in the community. It is also recommended that IBCs include: experts in biosafety and containment; persons knowledgeable in institutional policies and applicable laws; individuals reflecting community attitudes; and at least one representative member from the laboratory staff. Committee members cannot review a project in which they have been, or expect to be, involved or have a direct financial interest. Finally, the Guidelines provide that while opening IBC meetings to the public is suggested but not required, minutes of the meetings and submitted documents must be available to the public on request.

Because the NIH Guidelines require establishment of an IBC when research is conducted at or sponsored by an entity receiving any NIH support for rDNA research, even privately funded projects employing rDNA must adhere to the NIH Guidelines if they are being carried out at, or funded by, an organization that has any NIH contracts, grants, or other support for this kind of research. Additionally, some communities and real estate leases require compliance with the NIH Guidelines, making such compliance legally binding even for private companies. Adherence to the NIH Guidelines is mandatory and important because they stipulate biosafety and containment measures for rDNA research. Furthermore, they delineate critical ethical principles and outline key safety reporting requirements for human gene transfer research.

Most of the 400 or so IBCs registered with OBA are at institutions that are subject to the NIH Guidelines and for whom IBC registration is mandatory. While most of these institutions are academic, some industry-based IBCs are registered with NIH as a consequence of receiving NIH support for rDNA research (e.g., SBIR grants) and thereby becoming subject to the NIH Guidelines. In other instances, companies voluntarily comply with the NIH Guidelines as a means of observing the highest standards for safety practices; as part of that voluntary compliance, they register their IBCs with the NIH. Several federal agencies including the USDA and the Department of Veterans Affairs (VA) have made compliance with the NIH Guidelines a condition of their support of intramural and extramural research projects. Furthermore, a number of federal IBCs are registered with NIH, including those at the NIH, the Department of Energy (DOE) laboratories (including the Lawrence Livermore, Los Alamos, Oak Ridge, Sandia National Laboratories) and various VA medical centers and military research institutes such as the Uniformed Services University of Health Sciences, the Walter Reed Army Medical Center, and the U.S. Army's Medical Research Institute for Infectious Diseases. Some of these facilities are registered with NIH because they receive NIH support for their rDNA research and others because it is the policy of the department or agency to comply with the NIH Guidelines. The responsibility for the "enforcement" of the Guidelines is shared by the NIH Office of Biotechnology Activities, the Recombinant DNA Advisory Committee (RAC), IBCs at individual institutions, and by the principal investigators (PIs) themselves.[14]

FRAMEWORK FOR IMPLEMENTATION OF THE NIH GUIDELINES FOR RDNA RESEARCH

The Guidelines provide an administrative framework that specifies the roles and responsibilities of various federal officials, research institutions, and individual scientists. Significant responsibility is shared among the NIH Office of Biotechnology Activities (OBA), the RAC, IBCs at individual institutions, the principal investigator (PI), the Biological Safety Officer (BSO) at the institution, and by investigators themselves. Scientific advice on the technical aspects of risk assessment is provided by technical experts on the RAC; public input is provided by experts in nontechnical subjects and by the right of the public to comment on major actions.

The system is based upon a tiered set of reviews that encourages experimental design to be well thought out and provides a means for catching potential problems. The Guidelines distinguish among experiments: those needing approval of the IBC as well as the RAC and NIH director before initiation; those involving human testing that need approval of the IBC and the Institutional Review Boards (IRBs) as well as the RAC; those that require approval of the OBA and IBC; those that only require IBC approval; those that merely require notice to the IBC at the initiation of the experiment; and exempt experiments.[15]

The RAC is designated to consist of up to 21 voting members, including the chair. A majority of the voting members have to be knowledgeable in relevant scientific fields, such as molecular genetics, molecular biology, or rDNA research, including clinical gene transfer research. At least four members of the RAC have to be knowledgeable in fields such as public health, laboratory safety, occupational health, protection of human subjects of research, the environment, ethics, law, public attitudes, or related fields. Representatives of various federal agencies also serve as nonvoting members.[16] Over time, the degree of centralized federal oversight has been substantially reduced. Many of the central functions of the RAC have been delegated to IBCs.[17] Each institution (and the IBC acting on its behalf) has become responsible for ensuring that all rDNA research conducted at or sponsored by that institution is conducted in compliance with the NIH Guidelines.

The RAC is, however, still responsible for advising the NIH director on actions such as: (1) adopting changes in the NIH Guidelines; (2) assigning containment levels, changing containment levels, and approving experiments considered "Major Actions" under the NIH Guidelines; (3) promulgating and amending lists of classes of rDNA molecules to be exempt from the Guidelines because they do not present a significant risk to health or the environment; and (4) certifying new host vector systems.

The RAC is also responsible for: (1) identifying novel human gene transfer experiments deserving of public discussion; (2) transmitting to the NIH director specific comments/recommendations about human gene transfer experiments; (3) publicly reviewing human gene transfer clinical trial data and relevant information evaluated and summarized by the NIH OBA in accordance with the annual data reporting requirements; (4) identifying broad scientific, safety, social, and ethical issues relevant to gene therapy research as potential Gene Therapy Policy Conference topics; (5) identifying novel social, ethical, scientific, and safety issues relevant to specific human applications of gene transfer and providing the necessary guidance.

All institutions subject to the NIH Guidelines are required to establish and register an IBC for the review of rDNA research. The IBC is designed to provide a quasi-independent review of rDNA work done at an institution. It is responsible for: (1) reviewing all rDNA research conducted at or sponsored by the institution and approving those projects in conformity with the Guidelines; (2) periodically reviewing ongoing projects; (3) adopting emergency plans for spills and contamination; (4) lowering containment levels for certain rDNA and recombinant organisms in which the absence of harmful sequences has been established; and (5) reporting significant problems, violations, illnesses, or accidents to the NIH OBA.[18]

It is also the responsibility of the institution to appoint a Biological Safety Officer if it engages in large-scale research or production activities involving viable organisms containing rDNA molecules. If the institution engages in rDNA research at BL-3 or BL-4 (see below), the officer must be a member of the IBC. The officer's duties include: (1) conducting periodic inspections to ensure laboratory standards are rigorously followed; (2) reporting to the IBC and the institution any significant problems, violations of the Guidelines, and any significant research-related accidents or illnesses; (3) developing emergency plans for handling accidental spills and personnel contamination and investigating laboratory accidents involving rDNA research; (4) providing advice on laboratory security; and (5) providing technical advice to the PI and the IBC on research safety procedures.

Pre-initiation review of experiments by the RAC has been an important part of the oversight mechanism. Pre-initiation approval of experiments by NIH is required only for: (1) experiments that have not been assigned containment levels by the Guidelines; (2) experiments using new host-vector systems, which must be certified by NIH; (3) certain experiments requiring case-by-case approval; and (4) requests for exceptions from Guideline requirements. Prior to the initiation of these experiments the PI must submit a registration document to the IBC containing the following information: the source(s) of DNA; the nature of the inserted DNA sequences; the host(s) and vector(s) to be used; whether an attempt will be made to obtain expression of a foreign gene, and if so, the protein that will be produced; and the containment conditions that will be implemented as specified in the NIH Guidelines.

The initial RAC review process includes a determination as to whether the human gene transfer experiment presents characteristics that warrant public RAC review and discussion. The NIH OBA will notify the PI(s) about the results of the RAC's initial review. Two outcomes are possible: (1) the experiment does not present characteristics that warrant further review and discussion and is therefore exempt from public RAC review and discussion; or (2) the experiment presents characteristics that warrant public RAC review and discussion. Completion of the RAC review process is defined as: (1) receipt by the PI(s) of a letter from the NIH OBA indicating that the submission does not present characteristics that warrant public RAC review and discussion; or (2) receipt by the PI(s) of a letter from the NIH OBA after public RAC review that summarizes the committee's key comments and recommendations (if any).

TYPES OF EXPERIMENTS THAT REQUIRE IBC, RAC, AND NIH DIRECTOR REVIEW

At this time, only two categories of experiments are considered "major actions" that require decision by the NIH director after review by the IBC and the RAC. One category includes experiments that propose the "deliberate transfer of a drug resistance trait to microorganisms that are not known to acquire the trait naturally—if such acquisition could compromise the use of the drug to control disease agents in humans, veterinary medicine, or agriculture." The second category includes experiments that propose the deliberate formation of rDNA-containing genes for the biosynthesis of toxin molecules lethal for vertebrates at an LD_{50} of less than 100 nanograms per kilogram of body weight (e.g., microbial toxins such as the botulinum toxins, tetanus toxin, diphtheria toxin, and *Shigella dysenteriae* neurotoxin). The containment conditions or stipulation requirements for such experiments must be recommended by the RAC and set by NIH at the time of approval.

PHYSICAL AND BIOLOGICAL CONTAINMENT STRATEGIES FOR NIH-FUNDED rDNA RESEARCH ACTIVITIES

Regulated experiments must be carried out in accordance with physical and biological containment levels; the degree of containment is based upon the degree of potential hazard. Physical containment requires practices, equipment, and facility safeguards that lessen the chances that a recombinant organism might escape. As discussed above, the NIH first published safety guidelines in 1976, followed by the publication in 1984 of *Biosafety in Microbiological and Biomedical Laboratories* (BMBL).[19] The BMBL guidelines address laboratory safety procedures for working with and handling infectious disease agents—they do not address laboratory security issues. The BMBL categorizes infectious agents and laboratory activities into four classes or levels (BL-1 to BL-4) and establishes safety requirements for each level based upon risk. Factors considered in determining the level of containment include agent factors such as: virulence, pathogenicity, infectious dose, environmental stability, route of spread, communicability, operations, quantity, availability of vaccine or treatment, and gene product effects such as toxicity, physiological activity, and allergenicity. [20] Table 2-1 summarizes the major requirements for each of the BMBL biosafety levels.

Experiments involving levels 2 through 4 and restricted risk group host organisms require IBC approval before recombinant experiments can be conducted. At the highest level (BL-4), nothing that is created should have any possibility of escape or of coming in direct contact with any laboratory workers. The containment conditions or stipulation requirements for such experiments must be recommended by the RAC and set by NIH at the time of approval. Containment conditions for experiments involving the introduction of rDNA into restricted agents are set on a case-by-case basis following NIH OBA review. The recommended practices, safety equipment, and facility safeguards in these guidelines establish a code of practice that is complied with voluntarily, one that all members of a laboratory community can together embrace to safeguard their colleagues and to protect the public. A permit is also required for all facilities working with such agents, although clinical laboratories used for research, diagnostic, reference, and/or verification purposes need only be certified (but do not require a license).[21]

Some organisms, including smallpox (*Variola major*) may not be studied in the United States except at specified facilities. Smallpox is an acute contagious disease caused by *Variola* virus, a member of the orthopox virus family. It was one of the world's most feared diseases until it was eradicated by a collaborative global vaccination program led by the World Health Organization (WHO). The last known natural case was in Somalia in 1977. Smallpox was officially declared eradicated in 1980. All research activities, including storage of *Variola major* are restricted to two international collaborating centers for smallpox research. The WHO Collaborating Center for Smallpox Research[22] in the United States is located at the CDC in Atlanta, Georgia, the other is located at the VECTOR Laboratory in Koltsovo, Russia.

Since their initial appearance, the physical biocontainment levels for rDNA experiments have been progressively lowered over time. As experience provided confidence that rDNA technology could be applied without creating dangerous organisms that could not be contained, the prohi-

TABLE 2-1 Summary of Recommended Biosafety Levels (BSL) for Infectious Agents[1]

BSL	Agents	Practices	Safety Equipment (Primary Barriers)	Facilities (Secondary Barriers)
1	Not known to cause disease in healthy adults	Standard Microbiological Practices	None required	Open bench top sink required
2	Associated with human disease, hazard = auto-inoculation, ingestion, mucous membrane exposure	BSL-1 practice plus: - Limited access - Biohazard warning signs - 'Sharps' precautions - Biosafety manual defining any needed waste decontamination or medical surveillance policies	Class I or II BSCs or other physical containment devices used for all manipulations of agents that cause splashes or aerosols of infectious materials; PPEs: laboratory coats; gloves; face protection as needed	BSL-1 plus: Autoclave available
3	Indigenous or exotic agents with potential for aerosol trans-mission; disease may have serious or lethal consequences	BSL-2 practice plus: -Controlled access - Decontamination of lab clothing before laundering -Baseline serum	Class I or II BCSs or other physical containment devices used for all manipulations of agents; PPEs protective lab clothing; gloves; respiratory protection is needed	BSL-2 plus: - Physical separation from access corridors - Self-closing, double door access -Exhausted air not recirculated - Negative airflow into laboratory
4	Dangerous/exotic agents which pose high risk of life-threatening disease, aerosol-transmitted lab infections; or related agents with unknown risk of transmission	BSL-3 practices plus:-Clothing change before entering -Shower on exit -All material decontaminated on exit from facility	All procedures conducted in Class III BSCs or Class I or II BSCs in combination with full-body, air-supplied, positive pressure personnel suit	BSL-3 plus: - Separate building or isolated zone - Dedicated supply/exhaust, vacuum, and decon systems - Other require-ments outlined in the text

[1]From the CDC/NIH Biosafety Guidelines: Biosafety in Microbiological and Biomedical Laboratories.

bitions were replaced with a series of risk-based mechanisms for oversight and approval.

COMPLIANCE WITH AND ENFORCEMENT OF THE NIH GUIDELINES

The Principal Investigator (PI) is responsible for full compliance with the Guidelines in the conduct of rDNA research and for ensuring that the reporting requirements are fulfilled; the PI is held accountable for any reporting lapses. For experiments that require NIH approval prior to IBC approval, it is the responsibility of the PI to petition NIH OBA with the concurrence of the IBC.

Compliance with the Guidelines is accomplished by a combination of local self-regulation and limited federal oversight, with the ultimate enforcement resting in the federal funding power. Even if noncompliance were found, no penalties can be imposed other than restriction or termination of NIH funding. The primary mechanism in the Guidelines for enforcing compliance is local self-regulation. Noncompliance may result in: suspension, limitation, or termination of financial assistance for the noncompliant NIH-funded research project and of NIH funds for other rDNA research at the institution, or a requirement for prior NIH approval of any or all future rDNA projects at the institution.

The Guidelines are designed to encourage industry's voluntary compliance by creating a parallel system of project review and IBC approval analogous to that required for NIH-funded projects, modified to alleviate industry's concerns about protection of proprietary information. A company's IBC determines whether the facilities meet the standards for the large-scale containment level but only for information-gathering purposes rather than to enforce these guidelines assigned by the RAC. A working group of the RAC may visit the companies and their IBCs from time to time. An important provision here is a process whereby a corporation may request presubmission review of the records needed to register its projects with NIH. The HHS Freedom of Information Officer informally determines whether the records have to be released; if so, they are returned to the submitting company.[23]

REGULATION OF MICROBIAL AGENTS (LISTED AGENTS AND TOXINS)

The Antiterrorism and Effective Death Penalty Act of 1996 required the Secretary of HHS to establish and enforce safety procedures for the transfer of listed biological agents (select agents), including measures to ensure proper training and appropriate skills to handle such agents, and proper laboratory facilities to contain and dispose of such agents. These regulations provide:

- "safeguards to prevent access to listed biological agents for use in domestic or international terrorism or for any other criminal purpose
- procedures to protect public safety in the event of a transfer or potential transfer of a listed biological agent in violation of the established safety procedures and safeguards
- the appropriate availability of biological agents for research, education, and other legitimate purposes."[24]

The select agent list, which is subject to revision, includes those agents considered to be the greatest threats to human health. An expanded list of pathogens and toxins went into effect on February 11, 2003. Agricultural plant and animal pathogens are now also included; the other changes reflect taxonomic changes and a few reassessments of what constitutes the most dangerous biothreat agents.[25] The organisms and toxins covered by these regulations are also presented in Table 2-2.[26]

The PATRIOT Act makes it a criminal offense for any person to knowingly possess any biological agent, toxin, or delivery system of a type or in a quantity that, under the circumstances, is not reasonably justified by prophylactic, protective, bona fide research, or other peaceful purpose.[27] In addition, the new law prohibits transfer or possession of a listed biological agent or toxin by a "restricted person."[28]

Title II, Enhanced Controls of Dangerous Biological Agents and Toxins, of the Bioterrorism Response Act substantially broadens the regulatory obligations for laboratories working with select agents.[29] The Secretary of HHS has the authority to establish and enforce safety procedures,[30] including: (1) proper training and appropriate skills to handle such agents and toxins; (2) proper laboratory facilities to contain and dispose of such agents and toxins; (3) measures to prevent access to such agents and toxins for use in domestic or international terrorism or for any criminal purpose; (4) procedures to protect the public safety in the event of a violation of the safety or security measures; and (5) appropriate availability of biological agents and toxins for research, education, and other legitimate purposes.[31]

On February 7, 2003 the CDC's final interim rule, Possession, Use and Transfer of Select Agents, went into effect. On February 11, 2003 similar rules from the Animal and Plant Health Inspection Service (APHIS) of the USDA also went into effect. A USDA permit is required for work with plant or animal pathogens.[32] In accordance with accepted scientific and regulatory practices of the discipline of plant pathology, an exotic plant pathogen (e.g., virus, bacteria, or fungus) is one that is not known to occur

TABLE 2-2 Organisms and Toxins Covered by Regulations

1997 CDC (Transfer)[a]	2003 CDC (Possession)[b]	2003 USDA (Possession)[c]
		Human and Animal
Bacteria:	**Bacteria:**	**Health Agents:**
Bacillus anthracis	*Bacillus anthracis*	**Bacteria:**
Brucella abortus	*Brucella abortus*	*Bacillus anthracis*
B. meliterisis,	*B. melitensis,*	*Brucella abortus*
B. suis	*B. suis*	*Brucella melitensis*
Burkholderia (Pseudomonas)	*Burkholderia (Pseudomonas)*	*Brucella suis*
Mallei	*mallei*	*Burkholderia mallei*
Burkholderia (Pseudomonas)	*Burkholderia (Pseudomonas)*	*Burkholderia pseudomallei*
Pseudomallei	*pseudomallei*	*Clostridium botulinum*
Clostridium botulinum	*Clostridium botulinum*	*Clostridium perfringens*
Francisella tularensis	*Francisella tularensis*	epsilon toxin
Yersinia pestis	*Yersinia pestis*	*Francisella tularensis*
Exemptions: vaccine strains as described in Title 9 CFR, Part 78.1.	Exemptions: vaccine strains as described in Title 9 CFR, Part 78.1.	
Viruses:	**Viruses:**	**Viruses:**
Crimean-Congo hemorrhagic fever virus	Crimean-Congo hemorrhagic fever virus	Nipah virus
Eastern equine encephalitis virus	Eastern equine encephalitis virus	Eastern equine encephalitis virus
Ebola virus	Ebola viruses	Hendra virus
Lassa fever virus	Lassa fever virus	Rift Valley fever virus
Marburg virus	Marburg virus	Venezuelan equine encephalitis virus
Rift Valley fever virus	Rift Valley fever virus	
South American hemorrhagic fever viruses (Junin, Machupo, Sabia, Flexal, Guanarito)	South American hemorrhagic fever viruses (Junin, Machupo, Sabia, Flexal, Guanarito)	
Tick-borne encephalitis complex viruses	Tick-borne encephalitis complex viruses	
Variola major virus (smallpox virus)	Variola major virus (smallpox virus)	
Venezuelan equine enchepalitis virus	Venezuelan equine encephalitis virus	
Viruses causing hantavirus pulmonary syndrome	Viruses causing hantavirus pulmonary syndrome	
Yellow fever virus	Yellow fever virus	
Equine morbillivirus		

TABLE 2-2 Continued

1997 CDC (Transfer)[a]	2003 CDC (Possession)[b]	2003 USDA(Possession)[c]
Exemptions: Vaccine strains of viral agents (Junin virus strain candid #1, Rift Valley fever virus strain MP-12, Venezuelan Equine encephalitis virus strain TC-83, Yellow fever virus strain 17-D).	Exemptions: Vaccine strains of viral agents (Junin virus strain candid #1, Rift Valley fever virus strain MP-12, Venezuelan Equine encephalitis virus strain TC-83, Yellow fever virus strain 17-D).	
Toxins: Abrin Aflatoxins Botulinum toxins *Clostridium perfringens* epsilon toxin Conotoxins Diacetoxyscirpenol Ricin Saxitoxin Shigatoxin Staphylococcal enterotoxins Tetrodotoxin T-2 toxin	**Toxins:** Abrin Aflatoxins Botulinum toxins *Clostridium perfringens* epsilon toxin Conotoxins Diacetoxyscirpenol Ricin Saxitoxin Shigatoxin Staphylococcal enterotoxins Tetrodotoxin T-2 toxin	**Toxins:** Botulinum neurotoxins Botulinum neurotoxin producing species of *Clostridium* *Clostridium perfringens* epsilon toxin Shigatoxin Staphylococcal enterotoxins T-2 toxin
Exemptions: Toxins for medical use, inactivated for use as vaccines, or toxin preparations for biomedical research use at an LD50 for vertebrates of more than 100 ng/kg body weight are exempt. National standard toxins required for biologic potency testing as described in 9 CFR, Part 113 are exempt.	Exemptions: Toxins for medical use, inactivated for use as vaccines, or toxin preparations for biomedical research use at an LD50 for vertebrates of more than 100 ng/kg body weight are exempt. National standard toxins required for biologic potency testing as described in 9 CFR, Part 113 are exempt.	

continued on next page

TABLE 2-2 Continued

1997 CDC (Transfer)[a]	2003 CDC (Possession)[b]	2003 USDA(Possession)[c]
Rickettsiae:	**Rickettsiae:**	**Rickettsiae:**
Coxiella burnetii	*Coxiella burnetii*	*Coxiella burnetii*
Rickettsia prowazekii	*Rickettsia prowazekii*	
Rickettsia rickettsii	*Rickettsia rickettsii*	
Fungi:	**Fungi:**	**Fungi:**
Coccidioides immitis	*Coccidioides immitis*	*Coccidioides immitis*
		Animal Agents and Toxins:
		African horse sickness virus
		African swine fever virus
		Akabane virus
		Avian influenza virus (highly pathogenic)
		Bluetongue virus (exotic)
		Bovine spongiform encephalopathy agent
		Camel pox virus
		Classical swine fever virus
		Cowdria ruminantium (Heartwater)
		Foot-and-mouth disease virus
		Goat pox virus
		Japanese encephalitis virus
		Lumpy skin disease virus
		Malignant catarrbal fever virus (exotic)
		Menangle virus
		Mycoplasma capricolum/M. F38/M. Mycoides capri (contagious caprine pleuropneumonia)
		Mycoplasma mycoides mycoides (contagious bovine pleuro-pneumonia)
		Newcastle disease virus (VVND)
		Peste des petits ruminants virus

TABLE 2-2 Continued

1997 CDC (Transfer)[a]	2003 CDC (Possession)[b]	2003 USDA(Possession)[c]
		Rinderpest virus
		Sheep pox virus
		Swine vesicular disease virus
		Vesicular stomatitis virus (exotic)
		Plant Agents:
		Liberobacter africanus
		Liberobacter asiaticus
		Peronosclerospora philippinensis
		Phakopsora pachyrhizi
		Plum pox potyvirus
		Ralstonia solanacearum, race 3, biovar 2
		Sclerophthora rayssiae var. zeae
		Synchytrium endobioticum
		Xanthomonas oryzae pv. oryzicola
		Xylella fastidiosa (citrus Variegated chlorosis strain)

[a] Based on Title V (Nuclear, Biological, and Chemical Weapons Restrictions) of the Antiterrorism and Effective Death Penalty Act of 1996 (PL-104-132).
[b] Based on the 2002 CDC Select Agents list.
[c] Based on Federal Register Vol. 67, No. 240. Friday, December 13, 2003 Rules and Regulations.

within the United States. Determination of whether a pathogen has a potential for serious detrimental impact on managed (agricultural, forest, grassland) or natural ecosystems is made by the PI and the IBC, in consultation with scientists knowledgeable about plant diseases, crops, and ecosystems in the geographic area of the research.[33]

These regulations impose additional shipping and handling requirements on laboratory facilities that transfer or receive select agents capable of causing substantial harm to human health. They are designed to ensure that select agents are not shipped to parties who are not equipped to handle them properly or who lack proper authorization for their requests.

The major shift in the new regulations establishes who may possess select agents as well as who may send and receive those agents, adds biosecurity requirements to the biosafety requirements, incorporates the personnel restrictions of the PATRIOT Act, involves the FBI in performing background checks on individuals who may have access to or conduct research on select agents, and proscribes certain types of experiments.

POSSESSION OF SELECT AGENTS

The Antiterrorism and Effective Death Penalty Act of 1996 required the Secretary of HHS to issue regulations governing the transfer of biological agents that have the potential to pose a severe threat to public health and safety. The CDC was authorized to regulate transfers of pathogens of unique interest in terms of their capacity to be used as weapons (the select agents list).[34] Accordingly, the CDC required that laboratories transferring select agents be registered.[35]

The Bioterrorism Response Act[36] adds new requirements for the Secretary of HHS to consider in listing agents and in preventing unlawful access to agents during transfers.[37] Facilities that register their possession and use of listed agents and toxins must provide "information regarding the characterization of listed agents and toxins to facilitate their identification, including their source; and safeguard and security requirements for registered persons."[38] Regulations specified under this law must "include appropriate safeguard and security requirements for persons possessing, using, or transferring a listed agent or toxin commensurate with the risk such agent or toxin poses to public health and safety (including the risk of use in domestic or international terrorism)."[39] Registered facilities must limit access to listed biological agents and toxins only to those determined by the registered facility to have a legitimate need to handle or use select agents,[40] and the secretary must be notified if a listed agent is lost, stolen, or released outside a biocontainment area of a facility.[41]

IMPORTATION AND INTERSTATE SHIPMENT
OF ETIOLOGIC AGENTS

The importation or subsequent receipt of human pathogens and vectors of human disease is controlled by the Public Health Service Foreign Quarantine Regulations (42 CFR Part 71.156).[42] Packages containing human pathogens or vectors originating in foreign locations must have an importation permit issued by the CDC. The importer is legally responsible for ensuring that the foreign personnel package, label, and ship the infectious materials according to the Interstate Shipment of Etiological

Agents regulations (42 CFR Part 72), regulations of the Department of Transportation on Transportation of Etiologic Agents (49 CFR Part 173), and the Dangerous Goods Regulations of the International Air Transport Association. An applicant for a permit must be knowledgeable of safe practices and proficient in the handling of infectious materials, be directly responsible for work with the infectious materials, and reside at the receipt address for the facility where work with the material will occur. The permit application requires the importer to provide characterization information for the material, a description of the objectives of the intended use, and a designation of the biosafety level of the laboratory where the work will occur.

The CDC is also responsible for regulating the interstate shipment of indigenous human pathogens, diagnostic specimens, and biologic products. The shipment of these materials must be in compliance with the provisions of the Interstate Shipment of Etiological Agents regulations (42 CFR Part 72), which specify packaging and labeling requirements and procedures for notification of successful delivery or failure of delivery.

OVERSIGHT OF FOREIGN NATIONALS[43]

This section briefly describes the current and emerging system of granting permission, to non-U.S. citizens through the visa system, for both short-term and extended stays, as well as two of the tracking systems. Issues related to sharing information with non-U.S. citizens are addressed in Chapter 3. The system is still evolving, so any description of current practice runs the risk of becoming rapidly out-of-date.[44] At present, however, September 11th and its aftermath have significantly increased the level of scrutiny, the time involved, and the opacity of the process. It should also be noted that, beyond the requirements to designate responsible individuals in affected institutions, to date laws and regulations related to individuals have been almost entirely aimed at rejection and prevention. That is, they have been aimed to limit access rather than to create a process of licensing or certification that would convey some more general, authoritative approval for working in life sciences research comparable, for example, to the licenses doctors must obtain to practice medicine.

The September 11th terrorist attacks greatly increased the concern and accelerated the plans for improving efforts to provide adequate scrutiny of visa applications and to track foreign nationals once they entered the United States. Foreign scholars planning shorter visits are also affected by increased concern for security, with impacts on the ability of researchers to take part in international meetings, conferences, or international research collaborations. Over time, these various restrictions could potentially alter the way research is conducted and have the potential to impede scientific progress in the United States.

The Department of Homeland Security (DHS) has been given responsibility for the policy guidance and regulation governing the issuance of visas, with the Secretary for Homeland Security given ultimate authority to determine who may and who may not enter the United States. Where there are foreign policy considerations the Department of State will continue to exercise authority. Consular officers, who have responsibility for guiding the review and processing of visa applications, will also remain under the auspices of the Department of State. In testimony before the House Select Committee on Homeland Security on July 11, 2002, Secretary of State Colin L. Powell reported that the State Department adjudicated over 10 million nonimmigrant visa applications in 2001. Around 7.5 million visas, or about 70 percent of the total, were issued.[45]

Additional Security Checks on Visa Applications

The Visas Condor program, initiated in January 2002, seeks to identify terrorists by checking a visa applicant's name against various U.S. government databases. Applicants are also required to fill out additional forms and be interviewed, fingerprinted, and subjected to additional identifying measures and background checks. Those affected by the Visas Condor program are predominantly Muslim men between the ages of 16 and 45 who come from any of approximately 26 (mostly Islamic) countries, but the system also applies to countries such as Russia and China. The State Department's goal is eventually to have the Visas Condor process take less than ten business days.

In response to earlier concerns the State Department, in consultation with other federal agencies, had created a Technology Alert List to provide guidance about which areas of science and technology were of particular concern. Applications from individuals with expertise in one of these areas would be sent to Washington for further review, usually by an agency with expertise in that field and perhaps by the FBI or intelligence services. The 16 categories on the list include "chemical and biotechnology engineering," which covers "technologies associated with the development or production of biological and toxin agents, pathogenics, biological weapons research."[46] In practice, "technologies" tended to be defined broadly enough to affect life scientists doing a variety of research.

Since January 2002 the Visas Condor security checks and the Technology Alert List reviews have required explicit approval from Washington for each applicant. In the past, at least the Alert List review process permitted consular officers to issue visas if they had not received a negative report from Washington within a certain number of days, but that is no longer the case. The agencies that need to provide clearance are determined by the State Department's Bureau of Consular Affairs and include the CIA and the FBI, as well as any other agency with a potential interest in the applicant. All applicants must be positively cleared by all the agencies involved in the review. This led to the backlogs and time delays reported in recent months. Consular Affairs officers have reported that in 2002, they conducted 35,000 Visas Condor and about 14,000 other checks. While this represents about a threefold increase in the number of cases referred to Washington, D.C., it is nonetheless a very small percentage of the total number of cases.[47]

In May 2003, Secretary of State Powell announced additional requirements for those seeking nonimmigrant visas. Except for certain visa categories or for countries where a visa waiver is in effect, as of August 1, 2003 all individuals between the ages of 16 and 60 are required to undergo a personal interview as part of the visa application process.[48] Substantial delays and increasing backlogs are anticipated in the visa process, since no additional resources are being allocated to consular officers and no overtime is to be used to handle the additional interviews. Furthermore, a new legislative mandate also requires that, as of January 1, 2004, all visitors entering the United States on a visa will be photographed and fingerprinted as part of U.S.-VISIT, the enhanced security screening process.[49]

Tracking Systems

In addition to increased scrutiny of visa applications, the U.S. government is initiating a number of systems for tracking foreign students and visitors to the United States.

Student and Exchange Visitor Information System (SEVIS)

The new Student and Exchange Visitor Information System (SEVIS) is an electronic System aimed at keeping better track of foreign students once they have received visas to study in the United States.[50] The Bureau of Citizenship and Immigration Services (formerly the Immigration and Naturalization Service, which was incorporated into the Department of Homeland Security) is responsible for SEVIS, although the program was developed in cooperation with the Departments of State and Education. SEVIS is designed to collect and report data on international student or exchange visitor status and changes, such as a change in one's program of study. It also provides system alerts, event notifications, and basic reports to the end-user schools, programs, and INS field offices. The timetable for its implementation and for colleges and universities to come into compliance with its regulations was accelerated after September 11th. Schools wishing to accept foreign students were required to register with SEVIS by January 30, 2003.

Interagency Panel for Advanced Science and Security (IPASS)

IPASS is a response to an October 2001 Presidential Decision Directive, "Combating Terrorism Through Immigration Policies," which directed federal agencies to develop student immigration policies through which the country "prohibits certain students from receiving education and training in sensitive areas." The White House's Office of Science and Technology Policy (OSTP) has been working with the White House's Homeland Security Council and others to develop and implement IPASS, although at the time of this report the Executive Order to create IPASS had not yet been signed.

PROFESSIONAL EDUCATION AND RESPONSIBILITIES OF LIFE SCIENTISTS

The Center for the Study of Ethics in the Professions lists over 850 "codes of ethics" on its website. A code of ethics is that profession's contract with the society it serves establishing in often very general terms acceptable moral behavior for practitioners of that profession.[51] Some differ widely in their content, because of their origins and their specific purposes. Others are similar in the topics they cover and the general ethical standards they articulate, but differ in language and in the specific ethical problems or abuses they address.[52] The Annex at the end of this chapter presents a representative cross-section of medical and scientific codes of ethics, from the Hippocratic Oath to the American Society for Microbiology's code of ethics and ethical standards for society members.

There is a considerable literature on the formulation of professional oaths and codes of conduct. Some have called for the initiation of a pledge to be taken by scientists—perhaps at graduation—much as modernized versions of the Hippocratic Oath are taken by some medical students upon graduation.[53] Others focus less on the development of codes and more on the inclusion of an emphasis on the moral and social responsibilities of life scientists in the training of students and postdoctoral fellows. Particularly if efforts to address the social responsibilities of scientists are led by leaders in the field and senior investigators, it is argued, young scientists will come to value "the ethics of individual behavior within the scientific enterprise and the societal impact of scientific research."[54] Whether mandatory or voluntary, the adoption of codes of conduct by professional organizations or national academies of science, and the integration of ethics education into the training of students should serve to sensitize "young scientists to reflect on the wider consequences of their intended field of work."[55]

Arguably, to be effective, any policy or set of procedures intended to address concerns about the offensive application of life sciences research data will require "ownership" by the scientific community. To the extent that responsibilities to guard against intentional misuse are recognized in professional codes of conduct and explicated and examined in the context of the training of the next generation of practitioners in the life sciences, opportunities to develop and maintain ownership by the community will only be increased.

At the November 2001 Review Conference for the BWC, the United States formally proposed new ways to strengthen the regime against biological weapons. Among the recommendations put forward was one that called upon the countries that are parties to the BWC to support the development and adoption of a code of conduct for scientists working with pathogenic organisms. Among the guiding principles of such a code of conduct would be a statement that "scientists will use their knowledge and skills for the advancement of human welfare and will not conduct any activities directed toward the use of microorganisms or toxins for hostile purposes or in armed conflict."[56] This proposal will be taken up by an intersessional meeting of the parties to the BWC in 2005 in Geneva. Proposals for the creation of such professional codes of conduct for practitioners in the life sciences have also come from the International Committee of the Red Cross,[57] the U.K. Foreign and Commonwealth Office,[58] the European Union, The Royal Society,[59] and others. Other U.S.-proposed measures to strengthen national and international implementation of the BWC include the oversight of "high-risk" genetic engineering experiments, an issue that will be addressed in Chapter 4.

THE INTERNATIONAL SITUATION

There is a deep and long-standing foundation of scientific self-regulation, voluntary standards, and associational accreditation. Given the fundamentally international character of research in the life sciences, any serious attempt to prevent the misuse of research must include efforts at improving and harmonizing standards and practices internationally. Recently, this has been supplemented by some mandatory requirements on specific aspects of laboratory safety.

This section provides a brief overview of some of the major international programs. It also offers examples from the regulatory systems of two other countries with advanced biotechnology research capabilities: the United Kingdom and Japan. The Committee's charge did not extend to a comprehensive review of the international regulatory environment, but the

Committee did examine some of the existing systems for possible positive or negative examples that might be relevant to the evolving U.S. situation.

Laboratory Safety

The Organization for Economic Cooperation and Development (OECD) has drafted quality management requirements called the Good Laboratory Practice (GLP) guidelines. Many national governments require laboratories that carry out safety and toxicological testing for the approval of new products to meet the GLP guidelines.

Controls Over Access to and Transfers of Dangerous Pathogens

As discussed above, smallpox was declared to be eradicated by the WHO at the annual meeting of the World Health Assembly in May 1980. This led to the greatest international control over access to dangerous pathogens—an international agreement implemented by the WHO to restrict the repository of smallpox virus cultures to two designated facilities, one in the United States and one in Russia. All countries other than the United States and Russia were to destroy their remaining stocks. The WHO, however, had no enforcement authority or means of verification and relied entirely on the voluntary cooperation of member states, leaving uncertainties about compliance.[60]

There are several other nonlegally binding access control agreements. The Australia Group is an informal arrangement of 33 member countries plus the European Union that harmonizes national controls on the export of dual use materials and production equipment that, in the wrong hands, could increase the risk of assisting chemical and biological weapons (CBW) proliferation. The Group meets annually to discuss ways in which national-level export licensing measures can more effectively ensure that would-be "proliferators" are unable to obtain necessary inputs for CBW programs. Participants in the Australia Group do not undertake any legally binding obligations. By enhancing the effectiveness of national export licensing measures, the Australia Group's activities serve to support the objectives and purposes of the BWC. The participants in the Australia Group encourage all countries to take the necessary steps to ensure that they and their industries are not contributing to the spread of biological and chemical weapons. Export licensing measures demonstrate the determination of Australia Group countries to avoid involvement in the proliferation of these weapons in violation of international law and norms.[61] The effectiveness of the cooperation among the participating countries depends solely on their commitment to CBW nonproliferation goals and the effectiveness of the measures they each take on a national basis.

The European Union also imposes export controls on dual use biotechnology equipment and pathogenic microorganisms and toxins, including agents that could be used for biological warfare.[62] To date, it is the only regional organization to undertake such an effort.

International regulations apply far more comprehensively to transnational shipment of human, plant, and animal pathogens. Among the international organizations that set regulations controlling the international transfer of such material is the International Air Transport Association (IATA). IATA Dangerous Goods Regulations require that packaging used for the transport of materials in specified hazard groups meet defined standards.[63] Shippers of microorganisms within the more serious hazard groups must be trained by IATA certified and approved instructors. They also require shippers' declaration forms, which should accompany the package in duplicate, and specified labels are used for organisms in transit by air (IATA, 1998).[64]

The Universal Postal Union (UPU) has established strict regulations on the shipment of pathogens through the mail. Other organizations regulate specific modes of transport. These regulations are primarily directed to the prevention of accidental release, but they also operate to track (but not to limit) who is supplying and receiving pathogens. The European Union is the only regional organization to regulate the shipment of pathogens.[65]

The Situation in the United Kingdom

In light of the September 11th terrorist attacks in the United States, the focus has shifted from safety requirements in the laboratory toward greater scrutiny of dangerous substances and increasing the difficulty in gaining access to areas where such agents are stored and used. The Antiterrorism, Crime and Security Act 2001 (ATCSA) part VII, is instrumental in this approach and attempts to tighten controls on access to dangerous pathogens and toxins used in research establishments and laboratories in the U.K. The pathogens and toxins affected are specified in Schedule 5 of the Act classified with their ACDP hazard group.[66] "Dangerous substance" means: (a) anything that consists of or includes a substance for the time being mentioned in Schedule 5; or (b) anything that is infected with or otherwise carries any such substance. Further substances may be added to the list by order of the Secretary of State.

In addition, the ATCSA establishes the power to vet personnel working in such establishments and to mandate security provisions. The owner of any premises that possess or use a dangerous substance must notify the

Secretary of State. The occupier of the premises must ensure that only appropriate individuals are given access to the premises. A police officer may require provision of information about each person who has access to any dangerous substance kept or used there or who has access to specified premises and identify the access that the person has, or is proposed to have.

Moreover, a constable may require provision of information about what dangerous substances are kept or used at the premises, the measures taken to ensure the security of any such substance, and measures taken to ensure that access to the substance is given only to those whose activities require access and only in circumstances that ensure the security of the substance. A constable may require that measures be taken to ensure the security of any dangerous substance. To assess compliance with those measures, a constable may, after giving at least 2 days notice, enter any relevant premises at a reasonable time. A constable who has entered any premises may search the premises, building, or site; require any person who appears to the constable to be in charge of the premises, building, or site to facilitate any such inspection; and require any such person to answer any question.

If research establishments do not meet personnel or security requirements, access to dangerous pathogens and toxins could be withdrawn. Where the Secretary of State reasonably believes that adequate measures to ensure the security of any dangerous substance kept or used in any relevant premises are not being taken and are unlikely to be taken, he may give a direction to the occupier of the premises requiring him to dispose of the substance. Moreover, the Secretary of State may give directions to the occupier of any relevant premises requiring him to secure that the person identified in the directions is not to have access to any dangerous substance kept or used there nor to specified premises. The Secretary of State may not give the directions unless he believes that they are necessary in the interests of national security. Failure to comply with the relevant duties is punishable by imprisonment, a fine, or both.

Research Oversight

The United Kingdom

For research involving DNA, the U.K. has set up the Health and Safety Executive (HSE) under the Health and Safety at Work Act of 1974 (HWSA).[67] It is primarily concerned with the protection of human health from possible ill effects of any workplace activity. Genetic modification and any activities in which genetically modified cells or organisms are cultured, stored, used, transported, destroyed or disposed of, under conditions of containment, are subject to the control of HSE under the Genetically Modified Organisms (Contained Use) Regulations of 1992, which are made pursuant to the HSWA.

The following bodies were established specifically to provide policy guidance on issues arising from developments in modern biotechnology. The Advisory Committee on Genetic Modification (ACGM)[68] is a nonstatutory body that advises the Health and Safety Commission/Executive and Ministers on human and environmental safety of the contained use of genetically modified organisms under the GMO (Contained Use) Regulations 1992 (based on EC Directive 90/219), as amended. ACGM focuses on safety questions in the laboratory and industrial installations. It is not involved in policy approval. The Advisory Group on Scientific Advances in Genetics (AGSAG)[69] is a nonstatutory advisory body that advises the Chief Medical Officer and the Director of Research and Development (DH) on potential implications for public health and for the National Health Service (NHS) of scientific advances in genetics. It also advises the NHS executive board on innovative genetic services and their evaluation.

Japan

Guidelines for rDNA experimentation define basic conditions required to promote and ensure safety for rDNA and related experiments. The experiments must be conducted under proper safety measures generally employed in microbiological laboratories, incorporating combinations of physical and biological containment measures as required by the safety evaluation of the experiment. Large-scale experiments with genetically engineered organisms must be conducted in a facility that has appropriate containment measures. Laboratory workers must be aware of the necessity of safety measures in the experiments, actually take those measures, and must have been thoroughly trained to ensure their expertise in standard methods and practices in microbiological experiments.

CONCLUSIONS

International regulation of biology is complicated by the lack of a multilateral consensus as to the basic security framework to which controls can be consistently applied. In contrast, the International Atomic Energy Agency (IAEA) oversees the Nuclear Nonproliferation Treaty (NPT) and nuclear safeguards agreements are negotiated with member states on a bilateral basis. The Organization for the Prohibition of Chemical Weapons oversees implementation of the exceptionally detailed Chemical Weapons Convention. Nothing comparable exists with regard to the oversight of biotechnology.[70] The BWC articulates a widely shared global norm against the weaponization of pathogens[71] and establishes statutory but not regulatory obligations on parties to the Convention. Nor is there any international oversight organization for biology. Efforts to strengthen the BWC by adding provisions for verification and compliance foundered in 2001 on fundamental diplomatic differences of principle and in particular cost-benefit analyses as to the effectiveness of the measures being proposed and whether multilateral versus bilateral approaches were the best way to prevent the development of biological weapons.

Multilateral discussions are continuing on ways to strengthen compliance with this treaty. At the BWC review conference in November 2002, member states agreed on a U.S. proposal to hold intersessional meetings in each of the next three years (2003–2005) before the 2006 Review Conference to discuss the five voluntary measures put forward by the U.S. to strengthen the BWC.[72]

With regard to oversight of research, *no country* has developed guidelines and practices to address all aspects of biotechnology research. There are a range of norms, standards of conduct for research, regulations, and institutional practices, many of which have been developed to address questions about research involving human subjects or the treatment of laboratory animals. In addition, responsibility for regulation of various aspects of biotechnology research is frequently shared among different departments or agencies.

In the United States, the PATRIOT Act and the Bioterrorism Response Act already establish the statutory and regulatory basis for protecting biological materials from inadvertent misuse. Once fully implemented, the mandated registration for possession of select agents, designation of restricted individuals who may not possess select agents, and a regulatory system for the physical security of the most dangerous pathogens within the United States will provide a useful accounting of domestic laboratories engaged in legitimate research and some reduction in the risk of pathogens acquired from designated facilities falling into the hands of terrorists. The Committee stresses that implementation of current legislation must not be overly restrictive given the critical role that the development of effective vaccines, diagnostics, therapeutics, and detection systems, along with a responsive public health system, will play in providing protection against bioterrorism—and other serious health threats. Otherwise these legislative solutions may unintentionally limit the research on dangerous pathogens by legitimate laboratories and investigators. To be effective, a harmonized, international system for the regulatory oversight of the possession of dangerous pathogens and toxins, comparable to the one being put in place in the United States, is needed.

Moreover, the different regulations now on the books do not add up to a systematic, generally applicable, means for the United States to respond to the challenges posed by research in the life sciences employing advanced biotechnology methods. Nor do they address the issues surrounding how to "manage" the knowledge and technologies produced through these research activities. At the moment, "control" over the results of these "dual use" research activities may be implemented at the point of information dissemination in the peer-reviewed literature.[73] A critical question is whether the various regulations and laws can be adapted, enhanced, supplemented, and linked to provide a system of oversight that will give confidence that the potential risks of misuse of dual use research are being adequately addressed. The Committee's answer to that question is contained in the following chapters.

ANNEX

The Hippocratic Oath

I SWEAR by Apollo the physician and AEsculapius, and Health, and All-heal, and all the gods and goddesses, that, according to my ability and judgement, I will keep this Oath and this stipulation—to reckon him who taught me this Art equally dear to me as my parents, to share my substance with him, and relieve his necessities if required; to look upon his offspring in the same footing as my own brothers, and to teach them this art, if they shall wish to learn it, without fee or stipulation; and that by percept, lecture, and every other mode of instruction, I will impart a knowledge of the Art to my own sons, and those of my teachers, and to disciples bound by a stipulation and oath according to the law of medicine, but to none others. I will follow that system of regimen which, according to my ability and judgement, I consider for the benefit of my patients, and abstain from whatever is deleterious and mischievous. I will give no deadly medicine to any one if asked, nor suggest any such counsel; and in like manner I will not give to a woman a pessary to produce abortion. With purity and with holiness I will pass my life and practice my Art. I will not cut persons labouring under the stone, but will leave this to be done by men who are practitioners of this work. Into whatever houses I enter, I will go into them for the benefit of the sick, and will abstain from every voluntary act of mischief and corruption; and, further, from the seduction of females or males, or freemen and slaves. Whatever, in connection with my professional service, or not in connection with it, I see or hear, in the life of men, which ought not to be spoken of abroad, I will not divulge, as reckoning that all such should be kept secret. While I continue to keep this Oath unviolated, may it be granted to me to enjoy life and the practice of the art, respected by all men, in all times. But should I trespass and violate this Oath, may the reverse be my lot.

American Physical Society
Guidelines for Professional Conduct

The Constitution of the American Physical Society states that the objective of the Society shall be the advancement and diffusion of the knowledge of physics. It is the purpose of this statement to advance that objective by presenting ethical guidelines for Society members.

Each physicist is a citizen of the community of science. Each shares responsibility for the welfare of this community. Science is best advanced when there is mutual trust, based upon honest behavior, throughout the community. Acts of deception, or any other acts that deliberately compromise the advancement of science, are unacceptable. Honesty must be regarded as the cornerstone of ethics in science. Professional integrity in the formulation, conduct, and reporting of physics activities reflects not only on the reputations of individual physicists and their organizations, but also on the image and credibility of the physics profession as perceived by scientific colleagues, government and the public. It is important that the tradition of ethical behavior be carefully maintained and transmitted with enthusiasm to future generations.

Adopted November 10, 2002

American Chemical Society
The Chemist's Code of Conduct

The American Chemical Society expects its members to adhere to the highest ethical standards. Indeed, the federal Charter of the Society (1937) explicitly lists among its objectives "the improvement of the qualifications and usefulness of chemists through high standards of professional ethics, education, and attainments..."

Chemists have professional obligations to the public, to colleagues, and to science. One expression of these obligation is embodied in "The Chemist's Creed," approved by the ACS Council in 1965. The principles of conduct enumerated below are intended to replace "The Chemist's Creed." They were prepared by the Council Committee on Professional Relations, approved by the Council (March 16, 1994), and adopted by the Board of Directors (June 3, 1994) for the guidance of Society members in various professional dealings, especially those involving conflicts of interest.

Chemists Acknowledge Responsibilities To:

The Public

Chemists have a professional responsibility to serve the public interest and welfare and to further knowledge of science. Chemists should actively be concerned with the health and welfare of co-workers, consumers, and the community. Public comments on scientific matters should be made with care and precision, without unsubstantiated exaggerated, or premature statements.

The Science of Chemistry

Chemists should seek to advance chemical science, understand the limitations of their knowledge, and respect the truth. Chemists should ensure that their scientific contribution, and those of their collaborators are thorough, accurate, and unbiased in design, implementation, and presentation..

The Profession

Chemists should remain current with developments in their field, share ideas and information, keep accurate and complete laboratory records, maintain integrity in all conduct and publications, and give due credit to the contributions of others. Conflicts of interest and scientific misconduct, such as fabrication, and plagiarism, are incompatible with this Code.

The Employer

Chemists should promote and protect the legitimate interests of their employers, perform work honestly and competently, fulfill obligations, and safeguard proprietary information.

Employees

Chemists, as employers, should treat subordinates with respect for their professionalism and concern for their well-being, and provide them with a safe, congenial working environment, fair compensation, and proper acknowledgement of their scientific contributions.

Students

Chemists should regard the tutelage of students as trust conferred by society for the promotion of the student's learning and professional development. Each student should be treated respectfully and without exploitations.

Associates

Chemists should treat associates with respect, regardless of the level of their formal education, encourage them, learn with them, share ideas honestly, and give credit for their contributions.

Clients

Chemists should serve clients faithfully and incorruptibly, respect confidentiality, advise honestly, and charge fairly.

The Environment

Chemists should understand and anticipate the environmental consequences of their work. Chemists have responsiblity to avoid pollution
and to protect the environment.

American Society for Microbiology
Code of Ethics

(The Code of Ethics has been revised and approved by Council 2000)

The American Society for Microbiology is dedicated to the utilization of
microbiological sciences for the promotion of human welfare and for the
accumulation of knowledge. These goals demand honesty and truthfulness
in all activities sponsored or supported by the Society.

Ethics Standards for Society Members

Guiding Principles

(1) ASM members aim to uphold and advance the integrity and dignity of
the profession and practice of microbiology.

(2) ASM members aspire to use their knowledge and skills for the advancement of human welfare.

(3) ASM members are honest and impartial in their interactions with
their trainees, colleagues, employees, employers, clients, patients, and
the public.

(4) ASM members strive to increase the competence and prestige of the
profession and practice of microbiology by responsible action and by sharing the results of their research through academic and commercial endeavors, or public service.

(5) ASM members seek to maintain and expand their professional knowledge and skills.

Rules of Conduct

1. ASM members shall not commit scientific misconduct, defined as fabrication, falsification, or plagiarism. However, scientific error or incorrect
interpretation of research data that may occur as part of the scientific process does not constitute scientific misconduct.

2. ASM members shall avoid improper conflicts of interest and potential
abuse of privileged positions. ASM members shall make full disclosure of
financial and other interests that might present a conflict in ASM activities.

3. ASM members shall abide by the ASM standards of publication that are
contained in a document entitled "ASM Editorial Policies/Ethics: Procedures and Guidelines." The Instructions to Authors for each ASM journal
also articulate the ethical publication standards of the ASM. In regard to the
presentations made as annual ASM meetings, conferences and workshops,

the ethical standards that pertain to the publications of the Society will be observed.

4. ASM members shall take responsibility to report breaches of the Rules of Conduct and to recommend appropriate responses, as defined in the Ethics Review Process.

5. Members shall not represent any position as being that of the ASM unless it has approval of the appropriate unit of the ASM.

6. ASM members, by accepting membership in the Society, agree to abide by this Code of Ethics.

Student Pugwash Group
United States

I promise to work for a better world, where science and technology are used in socially responsible ways. I will not use my education for any purpose intended to harm human beings or the environment. Throughout my career, I will consider the ethical implications of my work before I take action. While the demands placed upon me may be great, I sign this declaration because I recognize that individual responsibility is the first step on the path to peace."

NOTES

[1] The P1-4 terminology, used to represent the four ascending levels of physical containment, was subsequently changed to correspond with the Biosafety Level 1-4 terminology later adopted by CDC and NIH.

[2] BMBL. 1984. HHS Publication No. (CDC) 86-8395; March. Also available from National Center for Injury Prevention and Control, Office of Health & Safety, Richmond, J.Y., et al. eds. 1999, "Biosafety in Microbiological and Biomedical Laboratories," 4th ed. HHS Publication No. (CDC) 93-8395, May.

[3] Richmond, J.Y. et al., eds. 1999. "Biosafety in Microbiological and Biomedical Laboratories," 4th edition, op. cit. p.5.

[4] The Biological Weapons Anti-Terrorism Act of 1989; 18 USC sec. 175.

[5] "The Antiterrorism and Effective Death Penalty Act of 1996," April 24, 42 U.S.C. 262 et seq. For a discussion of the events and considerations leading to this enactment, see Kellman, B. 2001. Biological Terrorism: Legal Measures for Preventing Catastrophe, Harvard Journal of Law and Public Policy 24:417.

[6] U.S. Congress. "Uniting and Strengthening America by Providing Appropriate Tools Required to Intercept and Obstruct Terrorism (USA Patriot Act) Act of 2001," Public Law 107-56, October 26. Available at http://news.findlaw.com/hdocs/docs/doj/oig71703patactrpt.pdf.

[7] U.S. Congress. "Public Health Security and Bioterrorism Preparedness and Response Act of 2002." P.L. 107-188. 42 U.S.C. 243, June 12. Available at http://tis.eh.doe.gov/biosafety/library/PL107-188.pdf.

[8] Sec. 817 of the Act concerns Expansion of the Biological Weapons Statute. A "restricted person" is defined as "anyone who: is under indictment for or has been convicted in any court of a crime punishable by imprisonment for a term exceeding one year; is a fugitive from justice; is an unlawful user of any controlled substance; is an alien illegally or unlawfully in the United States; has been adjudicated as a mental defective or has been committed to any mental institution; is an alien who is a national of a country which is currently designated by the Secretary of State as a supporter of terrorism; or has been dishonorably discharged from U.S. armed forces." Currently there are seven countries on the State Department's List of State Sponsors of Terrorism: Cuba; Libya; Iran; Iraq; North Korea; Sudan; and Syria.

[9] *Ibid.*

[10] In the Joint Explanatory Statement of the Committee of Conference, the Managers stated that the primary goals of the new provisions in the Law are to "ensure the prompt reporting to the Federal government of possession of select agents (including by those who were in possession prior to April 15, 1997, the effective date for reporting transfers of select agents), to increase the security over such agents (including access controls and screening of personnel), and to establish a comprehensive and detailed national database of the location and characterization of such agents and the identities of those in possession of them."

[11] See FBI Bioterrorism Preparedness Act: Entity/Individual Information Form at http://www.fbi.gov./terrorinfo/fd-961.pdf.

[12] For a discussion of the judiciary's role in overseeing protection of the public in this context, see Mack v. Califano, 447 F. Supp. 668 (D.D.C. 1978).

[13] More about IBCs and the registration process can be learned at the following website: http://www4.od.nih.gov/oba/IBC/IBCindexpg.htm.

[14] The RAC was first established in 1974, two years before the NIH Guidelines.

[15] Section III: Experiments Covered by the NIH Guidelines.

[16] Section IV-C-2.

[17] See Rosenblatt, D.P. 1982. "The Regulation of Recombinant DNA Research: The Alternative of Local Control." 10 *British Columbia Environmental Affairs Law Review* 37.

[18] Section IV-B-2-b: Functions of IBCs.

[19] The BMBL has issued instructions for laboratory directors to develop better methods of handling, storing, containing, and sterilizing infectious agents.

[20] Section II-A-3: Comprehensive Risk Assessment.

[21] See 42 U.S.C. s262 (a) 2000.

[22] Section V-L.

[23] Section IV-D-5-b: Pre-submission Review.

[24] U.S. Congress. Antiterrorism and Effective Death Penalty Act of 1996, P.L. 104-132, April 24, sec. 511.

[25] 42 CFR 73 for HHS; 7 CFR 331 and 9 CFR 121 for USDA.

[26] In determining whether to list a biological agent, the Secretary of HHS, in consultation with scientific experts representing appropriate professional groups, was required to consider the agent's effect on human health, its degree of contagiousness and methods by which the agent is transferred to humans, and the availability of immunizations and treatments for illnesses that may result from infection

by the agent. These regulations should include measures to ensure proper training and appropriate skills to handle such agents; and proper laboratory facilities to contain and dispose of such agents and provide: safeguards to prevent access to listed biological agents for use in domestic or international terrorism or for any other criminal purpose; procedures to protect public safety if there is a transfer or potential transfer of a listed biological agent violation of the established safety procedures and safeguards; and for the appropriate availability of biological agents for research, education, and other legitimate purposes.

[27] Neither the term "bona fide" nor "legitimate" is defined in the Act, however.

[28] See fn.8.

[29] Regulatory Control of Certain Biological Agents and Toxins, available at http://www.asmusa.org/pasrc/pl107188.pdf.

[30] The statute prohibits the knowing possession of any biological agent, toxin, or delivery system that is not reasonably justified for prophylactic, protective, bona fide research, or other peaceful purposes. In addition, the law makes it a criminal offense to allow restricted persons to possess, transport or receive select agents. U.S. Congress. Uniting and Strengthening America by Providing Appropriate Tools Required to Intercept and Obstruct Terrorism (USA Patriot Act) Act of 2001. P.L. 107-56, October 26, sec. 817.

[31] The Uniting and Strengthening America by Providing Appropriate Tools Required to Intercept and Obstruct Terrorism (USA PATRIOT Act) Act of 2001. P.L. 107-56.

[32] Section III-D-1-d.

[33] Section V-M.

[34] 42 biological agents and toxins are listed in Appendix A of 42 CFR Part 72.

[35] The purpose of registration was to control domestic transfers based upon a permitting system. A registered laboratory could legally transfer select agents only to another registered laboratory; some transfers were denied because of concerns about the adequacy of the facility proposed to receive the agent. Transfers to non-registered laboratories were prohibited. Registration, however, was principally a matter of notification: a laboratory was obligated to notify relevant authorities of a transfer to another registered facility and that the transfer itself complied with applicable safety standards. Specific information about particular pathogens that the facility possessed did not have to be reported, not even if they were the subjects of extensive research, so long as they were not transferred. This was not intended to be a strict licensing system but merely a way of overseeing transfers and shipments of lethal pathogens.

[36] 42 U.S.C. 243 *et seq.* New considerations for listing agents include the availability and effectiveness of pharmacotherapies as well as immunizations to treat and prevent any illness resulting from infection by the agent or toxin, the needs of children and other vulnerable populations, and consultations with groups with pediatric expertise. The secretary must establish and enforce safeguard and security measures to prevent access to listed biological agents and toxins for use in domestic or international terrorism or any other criminal purpose.

[37] The law further provides comparable regulatory authorities to the Secretary of the Department of Agriculture regarding the possession, use, or transfer of listed biological agents and toxins that present a severe threat to plant or animal health or animal or

plant products and includes provisions to facilitate coordination and cooperation between the Department of Agriculture and the Department of Health and Human Services with respect to agents or toxins that are regulated by both agencies.

[38] The Bioterrorism Response Act also establishes a national database to collect registration information including the names and locations of registered facilities; the listed biological agents and toxins they possess, use or transfer; and characterization and source data for listed agents they possess. The purpose of this database is to assist public health and law enforcement officials to identify the origin or source of a listed agent used to cause harm to the public.

[39] Persons (facilities) and individuals who possess, use, or transfer listed biological and toxins agents must register with the Secretary, Department of Health and Human Services. Registered facilities that transfer a select agent to any person one knows or has reasonable cause to believe has not registered could be fined or imprisoned up to five years or both. Also, whoever knowingly possesses a select agent for which the person has not obtained a registration shall be fined or imprisoned for up to five years.

[40] The Public Health, Security and Bioterrorism Preparedness Act, H.R. 3448, 107th Cong § 351A (e)(2)(A). Facilities should promptly submit the names of such individuals to the Secretary of Health and Human Services and the Attorney General who shall promptly use criminal, immigration, national security, and other electronic databases available to the federal government to check if the individual is a "restricted person."

[41] In the Joint Explanatory Statement of the Conference Committee's report, the Managers stated that the primary goals of the new provisions in the law are to "ensure the prompt reporting to the Federal government of possession of select agents (including by those who were in possession prior to April 15, 1997, the effective date for reporting transfers of select agents), to increase the security over such agents (including access controls and screening of personnel), and to establish a comprehensive and detailed national database of the location and characterization of such agents and the identities of those in possession of them."

[42] The Public Health Service Act was first passed in 1944, with numerous subsequent amendments to its provisions, including those governing the foreign quarantine regulations.

[43] A detailed account of the system in place as of early 2003, on which this section is based, may be found in White, W.D. and L. Peterson. 2003. "Visas for Visiting Students and Scientists: Current Situation (January)," The Physiologist, April. Available at http://www.the-aps.org/publications/tphys/2003html/april03/visas.htm.

[44] In response to continuing concerns about the impact of the visa system on international scientific collaboration and to the need to keep the scientific community informed about its responsibilities under the evolving system, The National Academies created an International Visitors Office in the spring of 2003. Its website may be found at www.nationalacademies.org/visas.

[45] Cited in White and Peterson, op. cit.

[46] Further information on the Technology Alert List and the screening system may be found on the State Department website, available at http://travel.state.gov/state147566.html.

[47] White and Peterson, op. cit. Available at http://www.the-aps.org/publications/tphys/2003html/april03/visas.htm.

[48] Unclassified State Department Cable 136100, May 21, 2003.

[49] October 26, 2004 is the deadline for biometrics on visas and passports.

[50] The State Department has designated seven countries as sponsors of terrorism: Iran, Iraq, Syria, Libya, Sudan, North Korea, and Cuba. An analysis of NSF data by Paula E. Stephan and her colleagues found that between 1990 and 1999 1,215 citizens from five of the seven countries received Ph.D.s from U.S. institutions (students from Cuba and North Korea were awarded fewer than five doctorates and were not included in the analysis). Of these, 147 were in fields that the authors considered "sensitive," including bacteriology, biochemistry, biotechnology research, microbiology, molecular biology, and neurosciences. Stephan, P.E., et al., "Survey of Foreign Recipients of U.S. Ph.D.s" Letter in *Science* 295 (5563):2211-2212.

[51] Parsons, P.J. 2001. "Ethics Codes: The Good, The Bad, and the Almost Ugly," PR Canada. Available at http://www.fastmpr.com/CODEST.HTM.

[52] Online Ethics Center: Codes of Ethics and Conduct, Online Ethics Center for Engineering and Science at Case Western Reserve University. Available at www.onlineethics.org; wysiwyg://17/http://onlineethics.org/codes/.

[53] Rotblat, J. 1999. "A Hippocratic Oath for Scientists," Editorial, *Science* 286 (5444): 1475. See also commentary by Kent, S. 1999. "Misuse of Science is Simply Wrong and the Scientists Involved are Responsible," *Science* 286 (5447) and Buckmaster, H. 2000. "Hippocratic Oath for Scientists," *Science*, February 8, 2000; and Gozum, M. 2000. "Societal Responsibilities," *Science* February 11, 2000.

[54] See S. Kent, op.cit.

[55] *Ibid.*

[56] U.S. Department of State. 2001. "New Ways to Strengthen the International Regime Against Biological Weapons," Fact Sheet, Bureau of Arms Control, Washington, D.C., October 19, p. 5.

[57] International Committee of the Red Cross. 2002. "Biotechnology, Weapons & Humanity: An Informal Meeting of Government and Independent Experts," Montreux, Switzerland, September 23-24.

[58] Foreign and Commonwealth Office. 2002. "Strengthening the Biological and Toxin Weapons Convention Countering the Threat from Biological Weapons," April 29. Presented to Parliament by the Secretary of State for Foreign and Commonwealth Affairs by Command of Her Majesty, April. Available at http://www.bradford.ac.uk/acad/sbtwc/other/fcobw.pdf.

[59] Joint statement by the Presidents of the National Academy of Sciences and The Royal Society. "Scientist Support for Biological Weapons Controls," *Science* 298 (5596):1135.

[60] For a detailed discussion of this issue see Tucker, J. 2001. *Scourge: The Once and Future Threat of Smallpox* (New York: Atlantic Monthly Press).

[61] Source: http://www.australiagroup.net/agbwc.htm. It should be noted, however, that some developing countries view the Australia Group as discriminatory and claim that it unfairly impedes the economic development of those states targeted by harmonized export controls.

[62] E.U. Council Regulation, 3381/94/EED. *On the Control of Export of Dual Use Goods* (Official J.L. 367), p. 1.

[63] IATA Dangerous Goods Regulation. 2004, 45th edition. IATA Packing Instruction 602 (Class 6.2) (IATA). For guidelines for shipping of microorganisms, see www.gbf.de/dsmz/shipping/shipping/htm.

[64] Smith, D., C. Rhode, and B. Holmes. *The Safe Handling and Distribution of Microorganisms under the Law*, at http://www.ukncc.co.uk/html/Information/docs/Postal.doc.

[65] September 1957. European Agreement concerning the International Carriage of Dangerous Goods by Road (ADR).

[66] The text of the Act is available at http://www.hmso.gov.uk/acts/acts2001/20010024.htm.

[67] Health & Safety Executive In Action. Available at http://www.hse.gov.uk.

[68] HSC Advisory Committee on Genetic Modification. Available at http://www.hse.gov.uk/foi/openacgm.htm.

[69] Human Genetics Advisory Commission Second Annual Report. 1999. Available at http://www.doh.gov.uk/hgac/papers/papers_f/f_09.htm.

[70] The CWC does include toxins, which provides a modest degree of international oversight for one portion of biology.

[71] Under Article IV of the BWC "[e]ach State party to this convention shall...take any necessary measures to prohibit and prevent the development, production, stockpiling, acquisition or retention of the agents, toxins, weapons, equipment and means of delivery specified in Article I of the Convention, within the territory of such State, under its jurisdiction or under its control anywhere."

[72] "Decision of the Fifth Review Conference of the Parties to the Convention on the Prohibition of the Development, Production, and Stockpiling of Bacteriological (Biological) Weapons and on Their Destruction," BWC/CONF.V/17, Geneva, November 2002, paras. 18-20. The first intersessional experts group meeting held in Geneva, Switzerland (August 2003) addressed regulation of pathogens by states parties to the BWC. Enhanced disease surveillance systems will be discussed in 2004 and "codes of conduct" in 2005.

[73] Journal Editors and Authors Group. 2003. "Statement on Scientific Publication and Security," *Science* 299 (5610):1149.

3

Information Restriction and Control Regimes

INTRODUCTION

The terrorist attacks of September 11, 2001 and the subsequent anthrax mailings in which five people died have produced a new sense of vulnerability in the United States. The governmental response has been wide ranging, affecting almost every sector of society. With respect to the life sciences, the most important initiatives to date are those embodied in the PATRIOT Act and the Bioterrorism Response Act. As discussed in Chapter 2, the latter legislation provides for the regulation of access to select agents and toxins through registration and screening of all institutions and individuals that possess, use, or transfer select agents.

Some have proposed that government control should go beyond the registration of laboratories and researchers who work with specified agents to include broad controls on the dissemination of research results, as well as the vetting of research proposals.[1] Should these proposals be adopted they would require a regulatory framework that would involve procedures for reviewing research proposals and restrictions on dissemination of research results; inevitably these regulations would profoundly affect research practices in biology laboratories. In effect, areas of life sciences research that were deemed "sensitive" because they could theoretically aid terrorists or be used in the production of biological weapons would be treated as secret.

Such a step should not be taken lightly; openness in science is highly valued. As the 1982 "Corson" report stated:

Free communication among scientists is viewed as an essential factor in scientific advance. Such communication enables critical new findings or new theories to be readily and systematically subjected to the scrutiny of others and thereby verified or debunked. Moreover, because science is a cumulative activity—each scientist builds on the work of others—the free availability of information both provides the foundations for further scientific advance and prevents needlessly redundant work. Such communications also serve to stimulate creativity, both because scientists compete keenly for the respect of their peers by attempting to be first in publishing the answers to difficult problems and because communication can inspire new lines of investigation. Finally, free communication helps to build the necessary willingness to confront any idea, no matter how eccentric, and to assess it on its merits.[2]

This chapter reviews the existing and emerging regulatory and oversight structure that governs the control of information related to biological research. Because issues of secrecy and sensitive information are new for much of the biological sciences, the chapter first discusses the experience of other scientific disciplines with these concerns. How other disciplines have addressed concerns about security suggests lessons that the Committee believes are relevant to the biological sciences as they respond to these issues.

PAST AS PROLOGUE?

The life sciences differ from the physical sciences in that they have not been deeply involved in developing new weapons in the United States since the Biological and Toxin Weapons Convention banned biological weapons in the early 1970s.[3] While many countries pursued BW work prior to the BWC entering into force, only a few had large-scale programs, and even in those countries military support for biological research was dwarfed by the resources going into nuclear and conventional weapons programs.[4]

The main patrons of research in the life sciences in the United States have been the National Institutes of Health (NIH), the Department of Agriculture (USDA), the National Science Foundation (NSF), the Department of Defense (DOD, the Department of Energy (DOE), and the pharmaceutical and agricultural industries. Secrecy issues involving national security concerns, as distinct from questions of intellectual property, have largely been absent in the life sciences. When the military did finance research in fields that had a biological content, such as oceanography, the result was to shift the balance in the field away from biology and toward the physical sciences that were familiar to the military program officers and advisory boards.[5] It should also be recognized, however, that the Defense Department has had a long-standing interest in fundamental basic and applied medical research for the development of diagnostic and medical countermeasures to "exotic" diseases that could adversely affect personnel readiness.

The Nuclear Weapons Complex

It is instructive to compare the situation in the life sciences to other areas of science in which the military has taken a stronger interest. The U.S. nuclear weapons program offers an example in which the Departments of Defense and Energy have played dominant roles in funding and shaping developments in nuclear physics and related fields. Nuclear weapons design is carried out in the Energy Department's national laboratories (primarily the Los Alamos National Laboratory, Sandia National Laboratories, and Lawrence Livermore National Laboratory). This work requires a special security clearance that restricts access to a small number of scientists who, in effect, constitute a closed society.[6] Although the principles that underlie the design of nuclear warheads are well understood by scientists around the world, the details of nuclear weapons design remain highly classified. Under the Atomic Energy Act of 1954, all information concerning nuclear weapons is "born classified," so that even research done outside the national laboratories under private sponsorship may be automatically classified if it is deemed relevant to nuclear weapons.[7] The category of "unclassified controlled nuclear information" (UCNI) is exempt from release under the Freedom of Information Act.[8] Exports of nuclear materials and related technologies are controlled under provisions of the Atomic Energy Act, the Nuclear Nonproliferation Act of 1978, and the Nuclear Nonproliferation Treaty, with nuclear dual use items covered by the Export Administration Act. As discussed below, "technology" can include information and various kinds of technical data and knowledge.

In short, there is a pervasive system of governmental secrecy and control for all research and development information related to nuclear weapons design and testing. Moreover, there is substantial consensus among scientists and the public that secrecy in the case of nuclear weapons is justified and should be maintained. Nuclear weapons scientists exchange their freedom to publish for relatively secure jobs and the satisfaction of feeling that they are contributing to national security.[9]

Other structural elements further distinguish nuclear weapons from biological and chemical weapons. Although some of the knowledge and facilities related to nuclear energy are relevant to the production of nuclear weapons, many steps intervene between the underlying science and successful production of a reliable nuclear weapon. Production of weapons-grade material, for example, requires industrial-scale processes. The proliferation of nuclear weapons capabilities has in every case so far required the resources of a state-sponsored program.[10] This means that in the case of nuclear weapons there are relatively few interests that might argue for greater openness for research results. By contrast, in the case of biological weapons, the basic science relevant for civilian uses is essentially the same as that relevant to military—and especially terrorist—applications. No "bright line" exists between purely defensive and purely offensive uses of infrastructure and knowledge. Box 3-1 offers additional comparisons of the distinct differences between fissile materials and biological pathogens that fundamentally affect the security concerns related to research relevant to nuclear and biological weapons.

Some scientists have argued that a small group of terrorists, using knowledge that has long been publicly available, could assemble a crude bomb based on highly enriched uranium (HEU). Such a device would not be a weapon of the sort designed in state weapons programs; it could, nevertheless, potentially equal the explosive power of the bomb used at

---

BOX 3-1
Characteristics of Fissile Materials and Pathogens

Fissile Materials	Biological Pathogens
Do not exist in nature	Generally found in nature
Nonliving, synthetic	Living, replicative
Difficult and costly to produce	Easy and cheap to produce
Not diverse: plutonium and highly enriched uranium are the only fissile materials used in nuclear weapons	Highly diverse: more than 20 pathogens are suitable for biological warfare
Can be inventoried and tracked in a quantitative manner	Because pathogens reproduce, inventory control is unreliable
Can be detected at a distance from the emission of ionizing radiation	Cannot be detected at a distance with available technologies
Weapons-grade fissile materials are stored at a limited number of military nuclear sites	Pathogens are present in many types of facilities and at multiple locations within a facility
Few nonmilitary applications (such as research reactors, thermo-electric generators, and production of radioisotopes)	Many legitimate applications in biomedical research and the pharmaceutical/biotechnology industry

Source: J.B. Tucker. 2003, "Preventing the Misuse of Pathogens: The Need for Global Biosecurity Standards," *Arms Control Today*, June.

Hiroshima. This possibility makes the safeguarding of stockpiles of fissile materials paramount, since controlling access to HEU remains the primary technical barrier to this type of weapons proliferation.[11]

Cryptography

Cryptography offers a second comparative case with potential lessons for the regulation of biotechnology. For hundreds of years cryptography was the province of governments that wanted to conceal state secrets—diplomatic and military—from others. To that end, they also kept information secret about the codes they used in order to make decryption by others less likely. The situation changed when private corporations developed a serious interest in cryptography as they began to do business electronically and needed to be able to conduct their affairs in private. Over the last three decades researchers in academic and industrial laboratories have entered the field in increasing numbers. Cryptography has thus become a dual use technology.

The new users of cryptography do not always see eye to eye with the government, which has an interest in retaining the ability to crack codes used by criminals and by foreigners doing business in the United States. In the 1990s controversy flared over the government's attempt as part of a new encryption standard to impose a key escrow feature, which would have enabled it to read any message it desired. The cryptography community and commercial interests argued that such a system would compromise privacy and undermine consumer confidence in business conducted over the Internet. It might also provide an opportunity for abuse of power by government agents holding the key. The encryption escrow feature was retained in the new standard, but use of the standard is voluntary.[12]

More significant for the debate over secrecy was the government's attempt in the late 1980s to put some academic research papers dealing with cryptography under export control laws and to require that research publications be vetted by the National Security Agency (NSA). The security concern was (and is) that open publication of some research could reveal vulnerabilities in encryption algorithms that an enemy could exploit. The cryptography community, by contrast, has argued consistently that in a well-designed cryptographic system only the key should be secret. In addition, algorithms should be public and open to challenge, so that problems can be quickly identified and fixed; that is, cryptography, like other sciences, progresses best under conditions of openness.

The outcome of this controversy provides an alternative model for addressing security concerns. In this case, the government dropped its efforts to impose secrecy in exchange for an informal system in which cryptographers often (but not always) submit their proposals and papers to the NSA for prepublication review, even when NSA is not the funding agency.[13] Thus, although there is no "born secret" category for cryptography research, the government has been able to keep track of ongoing research and to exercise some control over publication of results. Compliance is not universal, but there appears to be an informal norm in large segments of the cryptography community that cooperating with the government on this issue is a sign of good citizenship.

Lessons from the Comparisons

What lessons do these examples offer for possible governmental controls on research in the life sciences? We can compare structural conditions in the three cases in at least four dimensions. In all three cases, the government has an interest in controlling access to information for security reasons: there is a prima facie argument for keeping some research secret from potential enemies. The ease with which research results can be transformed into weapons or a technological advantage to be used against the United States varies sharply, however. In the case of nuclear weapons, understanding the principles behind the bomb is only part of what is needed to produce a weapon. Just as important—and far more difficult to obtain—is access to plutonium or weapons-grade uranium and the know-how to construct the device. Cryptography lies at the opposite extreme: a cryptographer can pursue his or her research with no more than pencil and paper, or at least with computing capabilities that are widely available.[14]

Research in biology lies between these two extremes. Traditionally, biology has been considered a small-scale science. Although work in genomics, proteomics, and bionanotechnology is overturning this paradigm, the research and development associated with biological weapons programs do not necessarily require large-scale investment or specialized, dedicated facilities. Creating pathogen weapons poses certain technical challenges, but the ability to produce enough material to cause morbidity, mortality, public panic, and economic costs is within the capability of many laboratories.

The degree to which the three technologies are dual use also varies. The civilian uses of nuclear energy have been cordoned off from weapons developments through a large investment in security classification, international diplomacy, and a discourse that insists that nuclear weapons are special.[15] Cryptography, as we have seen, has recently become a dual use technology, but its applications in the civilian world are growing rapidly in tandem with the Internet. The life sciences lie beyond cryptography in the dual use dimension; civilian uses dominate in the field, and the military's interest in the life sciences, although not negligible, has been dwarfed by their interest in the physical sciences. This is seen most clearly, perhaps, in the dominance until fairly recently of the NIH, NSF, and USDA, rather than DOD and DOE, in federal support of research in the life sciences.

The size of the field is also important when contemplating government controls. One reason the cryptography solution has worked relatively well may be that the number of publications is so small. The leading journal, *The Journal of Cryptology*, publishes only about 20 papers a year, and researchers present about 125 papers annually at conferences.[16] By contrast, the American Society for Microbiology's 11 journals publish 6,000 papers a year, and by some estimates there are between 10,000 and 20,000 journals published in the life sciences internationally.[17] Even if only a very small fraction of the research in these journals potentially arouses concern, the sheer volume of publication in the life sciences would make any effort to devise a screening mechanism for information deemed "sensitive"— or to ensure compliance—a daunting challenge.

In addition, there are many more life scientists than there are nuclear physicists or cryptographers. Only a few scientists are likely to be working with the list of select agents that have been the target of control so far, but many more would be included in a control regime that encompassed sensitive techniques as well as an expanded list of select agents. Furthermore, unlike the nuclear physicists, the life scientists are in many widely dispersed locations. The more people and facilities subject to control, the higher the costs.[18]

Finally, there is no established culture of working with the national security community among life scientists as currently exists in the fields of nuclear physics and cryptography. As a group, biologists lack the experience of either nuclear physicists or cryptographers in interacting with the security agencies of the federal government, and conversely those agencies lack close ties and working relationships with the life sciences community. The tradition of classified government research is well established in the latter two fields; the counterpart in the life sciences was the DOD program for research on biological weapons centered at Fort Detrick, MD, which ended in 1970. As noted above, however, that bioweapons program was only a small part of the government's funding of basic research in the life sciences and its very secrecy tended to isolate it from the larger community of life scientists. Since 1970, when President Nixon ended offensive biological weapons research, Fort Detrick has been a relatively open facility, housing a number of military and civilian tenants, including a large array of National Cancer Institute laboratories, as well as the U.S. Army Medical Research Institute for Infectious Diseases (USAMRIID). The number of life scientists involved in intramural research sponsored by the Department of the Army remains, however, relatively small.[19] It should be noted that other DOD agencies, such as the Defense Advanced Research Projects Agency (DARPA) and the Defense Threat Reduction Agency, have become significant sponsors of biodefense research, as have other federal agencies such as the Department of Homeland Security and the intelligence community. DARPA-sponsored research, while sometimes controversial, is unclassified.

The differences among the three areas are instructive. They suggest that controls on information flows in the life sciences will face obstacles rather different from those encountered in nuclear science and cryptography. The situation would be further complicated by the expansion of categories of information that the government wishes to control.

SECRET AND SENSITIVE INFORMATION

An excellent summary of the different types of information control regimes in the United States is published by the Association of American Universities and is reproduced in Box 3-2. [20]

Secret Information

The U.S. government handles issues of secrecy through a complex mix of statutes, regulations, and procedures that govern the control of classified information, public access to government information, and the

---

BOX 3-2
Definitions and Regulations Involved in the Classified-Sensitive Information-Unclassified Debate

Classified Research:
Executive Order 12958, issued on April 17, 1995, prescribes a uniform system for classifying, safeguarding, and declassifying national security information. Information may only be classified if certain conditions are met.

There are seven classification categories listed in section 1.5, the fifth of which is "scientific, technological, or economic matters relating to the national security." "National security" is defined as "the national defense of foreign relations of the United States."

Later, in section 1.8b, EO 12958 reiterates that basic scientific research information not clearly related to the national security may not be classified.

Classified projects are not published in open literature. Information is transferred only between those who obtain the required clearance. This applies even when the research is performed by scientists outside of government facilities.

Many universities do not accept classified projects. Many of those that do conduct research in facilities separate from the main campus.

Restricted data (RD), are classified according to a system created by the Atomic Energy Act of 1954 (AEA).[1]

The term "Restricted Data" means all data concerning (1) design, manufacture, or utilization of atomic weapons; (2) the production of special nuclear material; or (3) the use of special nuclear material in the production of energy, but shall not include data declassified or removed from the Restricted Data category pursuant to section 2162 of this title.[2]

The scope of the definition is broad and is rendered even more elastic by expansive definitions of "design" and of "research and development."[3] Unlike NSI, RD is interpreted by DOE as "born classified" — that is, to be considered a protected secret upon coming into existence without any affirmative act or decision by an official or, indeed, any involvement by government at all.[4] The AEA authorizes sealing off an entire area of scientific and engineering knowledge from public scrutiny. The AEA has provisions authorizing declassification of information falling within the scope of the definition.[5] Over the years, RD relating to many once-classified areas has been declassified, largely in order to facilitate commercial applications.[6] As a result, information relating to civil power reactors and nuclear fuel reprocessing is not classified. The remaining areas of national defense-related nuclear information that contain RD pertain to (1) nuclear weapon design; (2) nuclear material and nuclear weapon production; (3) certain theoretical aspects of inertial confinement fusion; (4) military reactors (production and submarine reactors); (5) isotope separation; and (6) directed nuclear energy systems.[7]

Sensitive Information Definitions

Sensitive Unclassified Information: The Computer Security Act of 1987 (P.L. 100-235) established requirements for protection of certain information on federal government automated information systems. This information is referred to as "sensitive" information, defined in the act as: "Any information the loss, misuse, or unauthorized access to or modification of which could adversely affect the national interest or the conduct of Federal programs or the privacy to which individuals are entitled under [the Privacy Act] but which has not been specifically authorized under criteria established by an Executive Order or an Act of Congress to be kept secret in the interest of national defense or foreign policy."

Sensitive But Unclassified: The Department of State describes "sensitive but unclassified" information as: "...information which warrants a degree of protection and administrative control that meets the criteria for exemption from public disclosure set forth under ... the Privacy Act." This is a document designation comparable to For Official Use Only. The Department of Defense also maintains several types of controlled, unclassified information but those too are similar to For Official Use Only.

Sensitive But Unclassified Technical Information: The Department of Energy's use of "sensitive but unclassified" is described as: "Information for which disclosure, misuse, alteration or destruction could adversely affect national security or government interests. National security interests are those unclassified matters that relate to the national defense or foreign relations of the Federal Government. Governmental interests are those related, but not limited, to the wide range of government or government-derived economic, human, financial, industrial, agricultural, technological, and law enforcement information, as well as the privacy or confidentiality of personal information provided to the Federal Government by its citizens."

Sensitive Homeland Security Information: OSTP Director Jack Marburger defined sensitive homeland security information during an October 10, 2002 appearance before the House Science Committee as "not a new category of information; rather it is the type of information that the government holds today which is not routinely available to the general public, such as law enforcement data and critical computer security threats or vulnerabilities."

Controlled But Unclassified: The Department of Defense has several categories of information called "controlled but unclassified."

Unclassified Research

NSDD-189, issued September 21, 1985, states the national policy for controlling the flow of science, technology, and engineering information produced in federally funded fundamental research at colleges, universities, and laboratories. NSDD-189 states, "to the maximum extent possible, the products of fundamental research remain unrestricted. It is also the policy of this Administration that, where the national security requires control, the mechanism for control of information generated during federally funded fundamental research in science, technology and engineering at colleges, universities and laboratories is classification. Each federal government agency is responsible for: a) determining whether classification is appropriate prior to the award of a research grant, contract, or cooperative agreement and, if so, controlling the research results through standard classification procedures; b) periodically reviewing all research grants, contracts or cooperative agreements for potential classification. No restriction may be placed upon the conduct or reporting of federally funded fundamental research that has not received national security classification, except as provided in applicable U.S. Statutes."

NSDD-189 defines fundamental research as: "basic and applied research in science and engineering, the results of which ordinarily are published and shared broadly within the scientific community, as distinguished from proprietary research and from industrial development, design, production, and product utilization, the results of which ordinarily are restricted for proprietary or national security reasons."

Policy Unchanged: NSDD-189 has not been superseded and continues to be the government policy. Assistant to the President for National Security Affairs Condoleeza Rice reaffirmed NSDD-189 on November 1, 2001 in a letter to Harold Brown, co-chairman of the Center for Strategic and International Studies. She stated, "this Administration will review and update as appropriate the export control policies that affect basic research in the United States. In the interim, the policy on the transfer of scientific, technical, and engineering information set forth in NSDD-189 shall remain in effect." OSTP Director Jack Marburger reaffirmed this position in a talk at the National Academy of Sciences on January 9, 2003.

Regulations: In the physical sciences, the distinction between what is harmful and what is not is relatively clear. This stems in part from the fact that those in the physical sciences have been dealing with these issues since World War II. It is more difficult to draw a distinction between knowledge that helps advance biomedical science and knowledge that can be used for deadly acts of bioterrorism. This makes it much more difficult to determine when and if information should be restricted. As a result, the regulations listed below focus mostly on the physical sciences, except for the last item, which is the newest.

Export Administration Regulation (EAR):

The Department of Commerce implements the EAR, which bars the export of items, technology, and technical information found on the Commerce Control List to foreign countries without appropriate export license. EAR covers the transfer of dual-use commercial goods. Dual-use technologies are those that have both a legitimate civilian and military use.

International Traffic in Arms Regulation (ITAR):

The Department of State implements the ITAR, which regulates the export of items on the Munitions Control List and technical information about them. Because technologies for space science are similar to those for military space applications, space scientists have encountered problems with exchange of items, information, and collaborations with foreign colleagues, students, and faculty. A March 2002 State Department change to ITAR attempted to ameliorate these problems by giving universities, in limited cases involving NATO and major non-NATO allies, an exemption for certain items and defense services based on 'public domain' information.

Both EAR and ITAR possess exemptions for "fundamental research." Both restate the NSDD-189 definition of fundamental research as ".. basic and applied research in science and engineering where the resulting information is ordinarily published and shared broadly within the scientific community," as distinguished from research the results of which are restricted for proprietary reasons or specific U.S. government access and dissemination controls. University research is not considered fundamental research if: (i) the University or its researchers accept other restrictions on publication of scientific and technical information resulting from the project or activity, or (ii) the research is funded by the U.S. government and specific access and dissemination controls protecting information resulting from the research are applicable.

A deemed export is transfer of information about controlled technologies to foreign nationals in the United States. Deemed exports may be regulated under the EAR (nondefense and dual-use technologies) or the Energy Department (information about special nuclear materials). ITAR refers to transfers of technical data to foreign nationals, whether in the US or abroad, as defense services.

Agency contract clauses:

Periodically, agencies insert new, restrictive language in contracts with universities. Most recently, restrictions on the participation of foreign nationals and/or on the disclosure of information have appeared in Department of Defense contracts. COGR has been compiling a list of these restrictions and is engaged in ongoing discussions with DOD and the Army about these clauses. The Army already has revised the new 4401 clause on release of Information once in response to university concerns, and is considering a further revision.

Since 1998, HHS regulations have restricted the transfer of certain biological agents and toxins ("select agents") to registered organizations, which included many universities. The select agent list consists of certain deadly viruses, bacteria, rickettsiae, fungi, and toxins and is determined by the Secretary of HHS. The USA PATRIOT Act (P.L. 107-56) prohibited possession of these agents, except by registered organizations, and barred access to these select agents by several classes of individuals, including those originating from countries sponsor terrorism. The Public Health Security and Bioterrorism Preparedness and Response Act of 2002 (P.L. 107-188) subsequently required institutions possessing select agents to improve security and access controls to the agents, develop a current inventory of those agents, and register their possession with HHS and USDA. Interim final regulations implementing P.L. 107-56 and 107-188 went into effect on February 11, 2003, and are found at 42 CFR 73. This is a new area of regulation and many of the processes and requirements are not yet clear.

---

[1] U.S. Code 42, "Congressional declaration of policy," § 2011 et seq. Available at http://www4.law.cornell.edu/uscode/42/2011.html?DB=uscode.

2 U.S. Code 42, "Definitions," § 2014 (y). Available at http://www4.law.cornell.edu/uscode/42/2014.html.

3 The AEA provides that "[t]he term 'design' means (1) specifications, plans, drawings, blueprints, and other items of the like nature; (2) the information contained therein; or (3) the research and development data pertinent to the information contained therein" (U.S. Code 42, "Definitions" § 2014 (i)). Available at http://www4.law.cornell.edu/uscode/42/2014.html. "Research and development" are defined as "(1) theoretical analysis, exploration, or experimentation; or (2) the extension of investigative findings and theories of a scientific or technical nature into practical application for experimental and demonstration purposes, including the experimental production and testing of models, devices, equipment, materials, and processes" (U.S. Code 42, "Definitions," § 2014 (x)). Available at http://www4.law.cornell.edu/uscode/42/2014.html. Neither definition is readily confined.

4 Hewlett, R.G. 1981. "Born Classified in the AEC: A Historian's View," *Bulletin of the Atomic Scientists* 37 (10), December: 20-27. Green, H.P. 1981. "Born Classified in the AEC: A Legal Perspective," *Bulletin of the Atomic Scientists* 37 (10), December: 28-30.

5 U.S. Code 42, "Classification and Declassification of Restricted Data," § 2162. Available at http://envirotext.eh.doe.gov/data/uscode/42/2162.html.

6 Some of the declassified information is still subject to control as unclassified controlled nuclear information (U.S. Code 42, "Applicability of Other Laws," § 2166. Available at http://www4.law.cornell.edu/uscode/42/2166.html).

7 Meridian Corporation. 1992. "Classification Policy Study," report prepared for the U.S. Department of Energy, Office of Classification, Washington, D.C., July 4:23.

maintenance of government records. Only those with security clearances are given access to classified information. Having a clearance, however, is not enough to provide access to classified information. An individual must also have a "need to know" the information in question. There is a formal system for controlling access to certain areas of information on a need-to-know basis. This additional layer of categories and controls adds to the complexity of the system.[21]

With two exceptions, only one of which is relevant to the life sciences, designating information as secret requires an affirmative action by a government official and can be applied only to information created within an agreed framework that makes classification a possibility. That is, the government has no authority to designate information produced outside this framework as classified; in effect, the classification system applies only to work done in government laboratories or under government contract. The first exception, described above, is the Atomic Energy Act, where information related to nuclear weapons may be "born classified" without any prior involvement of the government in its generation.

The second exception, which is potentially relevant to aspects of biotechnology research, permits information received as part of the patent application process to be classified. Under the Invention Secrecy Act of 1951, the government is required to impose "secrecy orders" on certain patent applications that contain sensitive information.[22] The disclosure of the invention is restricted and the grant of a patent is withheld. As summarized by the Project on Government Secrecy of the Federation of American Scientists, "[t]his requirement can be imposed even when the application is generated and entirely owned by a private individual or company. There are several types of secrecy order which range in severity from simple prohibitions on export (but allowing other disclosure for legitimate business purposes) to a classification requiring secure storage of the application and prohibition of all disclosure."[23]

In the wake of the September 11th attacks, President Bush has extended classification authority to a number of agencies not previously involved in these matters, including the USDA, HHS, and EPA. Their new authority will be exercised under the existing classification system.

Under the current system, in most agencies the task of managing potentially classified authority is so large that the authority to classify information has been delegated extensively. Literally thousands of government officials have classification authority. Detailed guides attempt to provide standards by which to judge whether particular information should be considered classified, but the more removed the issues are from specific details or facts, the more judgment becomes involved.[24] One would expect the same delegation to occur as the USDA, HHS and EPA exercise their new authority.

As discussed above, the struggle to decide whether areas of scientific research should be restricted in the name of national security recurred throughout the Cold War. During the early 1980s the Reagan Administration sought to restrict scientific communication in a number of fields. That controversy eventually led to a presidential directive in 1985, influenced in part by the Corson report.[25] National Security Decision Directive 189 (NSDD-189) states that federally funded fundamental research, such as that conducted in universities and laboratories, should "to the maximum extent possible" be unrestricted.[26] Where restriction is deemed necessary, the control mechanism is formal classification. "No restrictions may be placed upon the conduct or reporting of federally-funded fundamental research that has not received national security classification, except as provided in applicable U.S. statutes." This policy is still in force and was reaffirmed as recently as November 2001, pending completion of an administration review of export policy controls.[27]

For all the serious concerns that can arise over whether information is properly classified, and how such decisions are made, debates over the classification of scientific research take place within a system of reasonably well-specified and understood rules. For the life sciences, with the exception noted above of the potential imposition of secrecy in the patent process, the question of whether research would be carried out under the restrictions of classification and whether research results might be classified would be part of the initial process of defining the terms under which the research is funded. Far more problematic is the interest in designating certain areas of research and certain types of knowledge in the life sciences—wherever they are produced and however they are funded—as "sensitive but unclassified."

Sensitive Information

The issue of "sensitive information" is not new. Classification is only one of the ways in which the U.S. government controls public access to information. Across the federal government there are many other categories that apply narrowly or broadly to specific types of information.[28] Some of the categories are defined in statute, some through regulation, and some only through administrative practices. Different agencies may also assign a variety of civil and even criminal penalties for violation of their restrictions.[29]

The most extensive restrictions are exceptions to the Freedom of Information Act (FOIA), which enable the government to deny public access to particular classes of information.[30] The withheld information, considered "For Official Use Only," must be tied to a particular FOIA exemption, for example, to protect individual privacy or proprietary business information. Some of the categories of exemptions are sufficiently general, however, to give federal agencies considerable latitude in withholding information related to internal decision making.

At a time of heightened concern about the proliferation of weapons of mass destruction to states or terrorists, many kinds of information can seem potentially relevant to U.S. adversaries and hence targets for expanded controls. For scientists the chief concern in any government-imposed requirement to shield "sensitive" information lies in the potential fuzziness of the category, coupled with the severity of possible penalties for failing to protect the information.[31] The most relevant categories of restricted information for research in the biological sciences are those related to "sensitive but unclassified" information and to "dual use" information covered by export controls.

Defining "Sensitive" Information

Sensitive but unclassified (SBU) information includes information generated within the government but may also extend to knowledge generated purely in the private sector. Particularly in the wake of the September 11th attacks, the standard examples of sensitive information tend to be those that relate to the vulnerability of critical infrastructures, including facilities that are privately owned. The key nodes in an electricity grid are a frequently cited example, and information related to the design and operation of a nuclear power plant or transport of nuclear materials has long been protected as Unclassified Controlled Nuclear Information. In addition, it is easy to imagine that information such as the location of biological research programs might be considered sensitive, if the theft of select agents is considered a threat.[32] The Bioterrorism Response Act exempts information on possession of select agents from FOIA.

The Bush Administration has urged federal agencies to use all applicable exemptions to the Freedom of Information Act when considering requests for "sensitive but unclassified information." The White House assigned the Office of Management and Budget the task of developing uniform policy guidance for government agencies on defining and controlling sensitive information.[33] Section 892 of the Homeland Security Act directs the President to "identify and safeguard homeland security information that is sensitive but unclassified."[34] No definition of "sensitive" is provided in the statute, however, and the fact that different agencies have put forward different definitions is a further concern.[35] A key question is whether restraints on sensitive information might be extended beyond the information held internally by federal agencies, and, if so, who would be responsible for determining what counts as "sensitive."

The DOE provides an illustrative example of the difficulties associated with attempts to define "sensitive" information. In the wake of the scandals over alleged Chinese spying at DOE laboratories in 1999, the Defense Authorization Act for FY2000 added a provision that imposed signficant civil penalties for disclosure of "sensitive" information even though no implementing regulations were ever produced. In January 2000 the DOE general counsel took the position that, since no definition of sensitive information existed in the Atomic Energy Act or departmental regulations, legal restrictions could not be applied or enforced on DOE employees or federal contractors.[36]

More generally, one basic DOE document defined "sensitive but unclassified information" as:

> Information for which disclosure, misuse, alteration, or destruction could adversely affect national security or government interests. National se-

curity interests are those unclassified matters that relate to the national defense or foreign relations of the Federal Government. Government interests are those related, but not limited to, the wide range of government or government-derived economic, human, financial, industrial, agricultural, technological, and law enforcement information, as well as the privacy or confidentiality of information provided to the Federal Government by its citizens.[37]

DOE also maintains a "Sensitive Subjects List" to guide what information can be discussed with non-U.S. citizens without triggering the need for an export license (see below). One list promulgated in 1999 included a set of topics under "Chemical and Biological Weapons" described as "illustrative but not exhaustive." Information regarding these topics "may pertain to the research, design, development, testing, manufacturing, production, or use." The proposed topics included (numbering from original):

2. Genetic research, techniques, and specialized equipment related to chemical or biological agents, *for example*:
a) genome sequences and databases
b) genetic engineering techniques

7. Defenses against, and vulnerabilities to, the use of chemical or biological agents, *for example*:
a) vaccines, antitoxins
b) equipment, including protective clothing[38]

This example is directly relevant only to DOE or DOE-contractor employees—although important work on the human genome is conducted at DOE facilities. Nonetheless, it illustrates the dilemmas scientists could face if there were a concerted government effort to promulgate and enforce such categories.

Recently, however, DOE has sought to clarify its policies, in part to address concerns about whether real or anticipated restrictions were hampering the vitality of the national laboratories. The Department is attempting to implement a policy in which "Official Use Only" information, based on the Freedom of Information Act, will be the standard for deciding whether and how to control unclassified information, gradually replacing the various other information categories that have emerged over the years.[39] In addition, a department-wide memo on May 12, 2003 reaffirmed that the provisions of NSDD-189 remain the basis for DOE policy regarding restrictions on fundamental research.[40] These DOE policies would presumably need to be reconciled with the regulations that will be required to define and implement Section 892 of the Homeland Security Act, which created the new category of "sensitive homeland security information."

Dual Use Information

Scientists may also face restrictions on their communications with foreign colleagues under various export control restrictions on sharing information regarding dual use technology. These restrictions can apply to communication both within the United States and with scientists abroad. Limits on foreign scientists through the visa system were described in Chapter 2. As with the Technology Alert List designed to prompt scrutiny of visa applications, the export controls governed by the Export Administration Act and its implementing regulations also extend to the transfer of dual use technology. Technology is considered "specific information necessary for the 'development,' 'production,' or 'use' of a product," and providing such information to a foreign national within the United States may be considered a "deemed export" whose transfer requires an export license.[41] Technology "which arises during or as a result of fundamental research" is not subject to export restrictions—which relieves many scientists—but not those engaged in proprietary research sponsored by commercial interests at public and private universities.[42]

PUBLICATION OF SENSITIVE INFORMATION IN THE LIFE SCIENCES

Until recently, there were very few cases of problems related to the publication of research results in the life sciences that attracted significant public attention. Some specialists in bioterrorism, however, had warned that, given continuing advances in biotechnology, open publication could provide information of use to terrorists.[43] The publication of the "mousepox" study, as well as other studies discussed in Chapter 1, made the issue a major concern for journal editors.[44] The public perception of potential risks associated with publication of such information led to calls for scientific journals to refrain from publishing "dangerous" research or to delete some data from published research results in order to preclude others from replicating the results.[45] Journals in the life sciences have responded in a number of ways to the concerns that published articles might provide useful knowledge or a road map for terrorists or rogue states.

In addition to the results of fundamental research, the compilation, synthesis, and assessment of already published results in review articles may provide an understanding of a field that could guide or assist terrorists. Even more difficult are the concerns raised by reports that result when scientists are assembled to render their judgment as experts about particular problems, even when they rely completely on open sources of information.[46] Against these risks, one must weigh the genuine service to the research community provided by review articles and the contributions of expert panels to informed public debate and decision-making on issues where scientific knowledge and perspective play a role. The Committee wanted to acknowledge these problems, which it expects will remain and perhaps grow as a concern, but they are beyond the scope of this report.

In response to the concerns about publication of research results, the American Society for Microbiology (ASM) determined that the 11 journals it publishes would not restrict the information in the materials and methods section of articles. But ASM has also instituted formal procedures as part of the peer-review process for submitted articles so that reviewers address the potential risks of the research results to national security. At present, these policies apply primarily—although not exclusively—to research conducted on select agents.[47] In 2002, of the 13,929 manuscripts submitted to ASM journals, 313 select agent manuscripts received special screening, and of these two manuscripts received additional screening by the full ASM publication board. The statistics through July 2003 are 8,557 manuscripts submitted, 262 select agent manuscripts screened, and none referred to the publication board for further review.[48] Other journals, such as *Science*, the *Proceedings of the National Academy of Sciences*, and *Nature*, have also become alert to potential articles that could cause concern and have moved to develop review procedures of their own.

These new procedures have been the subject of intense discussion within the life sciences community and between life scientists and the national security community. In January 2003, for example, the National Academy of Sciences and the Center for Strategic and International Studies convened a one-day workshop to review the general question of sensitive information and specific issues of publication. Gatherings such as this attest to the seriousness with which the scientific and national security communities regard these issues but also to the difficulty of establishing productive communication—and even more of devising satisfactory, workable solutions.[49]

In mid-February 2003, the editors of the major journals in the life sciences, including *Nature*, *Cell*, *Science*, and the *Proceedings of the National Academy of Sciences (PNAS)*, published a joint statement on "Scientific Publication and Security."[50] The statement, which appears in Box 3-3, was the outcome of discussions begun at the January workshop. It has generated substantial comment and controversy, but is also an example of the efforts of the scientific community to respond to issues related to potential risks through the development of self-governance mechanisms.[51]

BOX 3-3
Statement on Scientific Publication and Security

PREAMBLE

The process of scientific publication, through which new findings are reviewed for quality and then presented to the rest of the scientific community and the public, is a vital element in our national life. New discoveries reported in research papers have helped improve the human condition in myriad ways: protecting public health, multiplying agricultural yields, fostering technological development and economic growth, and enhancing global stability and security.

But new science, as we know, may sometimes have costs as well as benefits. The prospect that weapons of mass destruction might find their way into the hands of terrorists did not suddenly appear on September 11, 2001. A policy focus on nuclear proliferation, no stranger to the physics community, has been with us for many years. But the events of September 11 brought a new understanding of the urgency of dealing with terrorism. And the subsequent harmful use of infectious agents brought a new set of issues to the life sciences. As a result, questions have been asked by the scientists themselves and by some political leaders about the possibility that new information published in research journals might give aid to those with malevolent ends.

Journals that dealt especially with microbiology, infectious agents, public health and plant and agricultural systems faced these issues earlier than some others, and have attempted to deal with them. The American Society for Microbiology, in particular, urged the National Academy of Sciences to take an active role in organizing a meeting of publishers, scientists, security experts and government officials to explore the issues and discuss what steps might be taken to resolve them. In a one-day workshop at the Academy in Washington co-hosted by the Center for Strategic and International Studies on January 9, 2003, an open forum was held for that purpose. A day later, a group of journal editors, augmented by scientist-authors, government officials and others, held a separate meeting designed to explore possible approaches.

What follows reflects some outcomes of that preliminary discussion. Fundamental is a view, shared by nearly all, that there is information that, although we cannot now capture it with lists or definitions, presents enough risk of use by terrorists that it should not be published. How and by what processes it might be identified will continue to challenge us, because—as all present acknowledged—it is also true that open publication brings benefits not only to public health but also in efforts to combat terrorism.

The statements follow:

FIRST: The scientific information published in peer-reviewed research journals carries special status, and confers unique responsibilities on editors and authors. We must protect the integrity of the scientific process by publishing manuscripts of high quality, in sufficient detail to permit reproducibility. Without independent verification—a requirement for scientific progress—we can neither advance biomedical research nor provide the knowledge base for building strong biodefense systems.

SECOND: We recognize that the prospect of bioterrorism has raised legitimate concerns about the potential abuse of published information, but also recognize that research in the very same fields will be critical to society in meeting the challenges of defense. We are committed to dealing responsibly and effectively with safety and security issues that may be raised by papers submitted for publication, and to increasing our capacity to identify such issues as they arise.

THIRD: Scientists and their journals should consider the appropriate level and design of processes to accomplish effective review of papers that raise such security issues. Journals in disciplines that have attracted numbers of such papers have already devised procedures that might be employed as models in considering process design. Some of us represent some of those journals; others among us are committed to the timely implementation of such processes, about which we will notify our readers and authors.

FOURTH: We recognize that on occasions an editor may conclude that the potential harm of publication outweighs the potential societal benefits. Under such circumstances, the paper should be modified, or not be published. Scientific information is also communicated by other means: seminars, meetings, electronic posting, etc. Journals and scientific societies can play an important role in encouraging investigators to communicate results of research in ways that maximize public benefits and minimize risks of misuse.

CONCLUSIONS

Any argument about imposing information controls—whether through formal classification or restrictions on "sensitive" information—must be made in the context of the specific institutional history and research culture of the life sciences research community. Like all sciences, the life sciences rely upon a culture of openness in research, where the free exchange of ideas allows researchers to build on the results of others, while simultaneously opening scientific results to critical scrutiny so that mistakes can be recognized and corrected sooner rather than later. Most scientists would argue that the openness that characterizes much of the scientific research enterprise is the source of the extraordinary gains in scientific knowledge that have enriched us materially and intellectually.

It is not that individual researchers, research groups, university administrators, or editors do not know how to keep secrets. Anyone who has spent much time in a university recognizes that there are categories of information that are not widely shared, from faculty salaries at private universities to the location of animal testing facilities. Academic journals and funding agencies keep secret the names of their reviewers. Research performed under contract to proprietary interests routinely requires a period of secrecy and prepublication review of manuscripts intended for presentation and publication in the peer-reviewed literature as a contractual condition of funding. These areas of secrecy, however, are the result of widely accepted understandings and local negotiations, and the number of people affected is limited to those directly concerned.

As already discussed, compared with other disciplines such as physics, the life sciences have relatively little experience with classified research. Beyond this, the life sciences cover a broad set of disciplines, from evolution and ecology to genomics and proteomics. Unlike nuclear weapons research, much of life sciences research is not of interest either to "rogue" offensive weapons programs or to potential terrorists. The range of scientists and institutions affected would thus be hard to enumerate, let alone monitor.

The costs of complying with information controls on life sciences research would range from their impact on the culture of the research laboratories, which is generally acknowledged to be extraordinarily open, to financial costs borne by institutions in complying with government regulations, to the creation of obstacles to monitoring compliance with international arms control measures directed at biological weapons. The restrictions already in effect on select agents have caused some laboratories to destroy archived samples and to limit exchanges of materials between scientists. To extend government controls to the information contained in laboratory reports, conference papers, and journal articles would further constrict avenues of communication, both formal and informal, which have been an essential source of the dynamism of biological research in the modern era.

Perhaps most important, major universities have proscribed classified research on campus. Those who do accept classified research have usually created separate facilities where access can be limited and controlled.[52] Secrecy would thus deprive the government of the graduate students and postdoctoral fellows who drive much of biological research—in many cases the best minds engaged in rapidly developing fields. Even without formal classification, the specter of information controls on "sensitive" information, given the current vagueness of the categories and the great difficulty of being any more exact about most of the dual use research, could be a significant deterrent to scientists to undertake research in some areas, such as infectious diseases. Yet these are precisely the areas where the best researchers are needed to help develop the nation's defenses against biological weapons, bioterrorism, and emerging-disease threats.

Thus there is a danger that the life sciences as a field of study would come to be regarded as less inviting, affecting the quality of researchers entering the field or making it more attractive to work outside the United States. Unlike the situation with nuclear weapons design/development/production and testing, biotechnology-related research in the life sciences is an international activity and proliferation-relevant knowledge is widely held. Limiting the development of biotechnology in the United States would reduce our worldwide competitiveness in this rapidly changing field. We conclude that imposing mandatory information controls on research in the life sciences, if attempted, will be difficult and expensive with little likely gain in genuine security. The next chapter describes the system that the Committee has concluded can best meet the needs of reducing the risks of misuse of biological research while still enabling vitally needed research to meet civilian and biodefense needs to go forward.

NOTES

[1] The Department of Defense, for example, proposed in early 2002 that researchers be required "to obtain DoD approval to discuss or publish findings of all military-sponsored unclassified research." The draft was withdrawn in the face of considerable criticism from the research community. Knezo, G.J. 2003. "Sensitive But Unclassified' and Other Federal Security Controls on Scientific and Technical Information: History and Current Controversy." (Washington, D.C.: Congressional Research Service), April 2. A critique of the proposed regulations by Don DeYoung, executive assistant to the director of research at the U.S. Naval Research Laboratory, can be found at: www.fas.org/sgp/othergov/deyoung.html.

[2] National Academy of Sciences. 1982. *Scientific Communication and National Security.* Washington, D.C.: National Academy Press. Available at www. nap.edu/books/0309033322/html/, p. 24. This is often called the "Corson report," after its chair, Dale Corson, president emeritus of Cornell University.

[3] The 1925 Geneva Protocol had already banned the "use in war" of chemical and biological weapons. The 1972 Biological and Toxin Weapons Convention, which entered into force in 1975, bans the development, production, stockpiling, and transfer of biological weapons.

[4] For a description of national chemical and biological weapons programs prior to 1970, see SIPRI. 1973. "The Problem of Chemical and Biological Warfare," *CB*

Weapons Today, Vol. II (New York: Humanities Press), Chap. 3. As noted in Chapter 1, the Soviet Union maintained a secret BW program into the 1990s.

[5] Doel, R. 2001. "Constituting the postwar earth sciences: The military's influence on the environmental sciences in America after 1945." *Social Studies of Science* 33:5 [Special Issue: The Earth Sciences in the Cold War, J. Cloud and J. Reppy, guest editors].

[6] Gusterson, H. 1996. *Nuclear Rites: A Weapons Laboratory at the End of the Cold War* (Berkeley: University of California Press), Chapter 4.

[7] Hewlett, R.G. 1981. "Born classified in the AEC: A historian's view." *Bulletin of the Atomic Scientist*, 37, December:20-27.

[8] In 1981 Congress gave the Department of Energy authority under the AEA to prevent unauthorized dissemination of information regarding "unclassified controlled nuclear information" (UCNI), which includes: (1) the design of production or utilization facilities; (2) security measures for such facilities or for nuclear material in such facilities or in transit; and (3) the "design, manufacture, or utilization of any atomic weapon or component if that information has previously been declassified or removed from the restricted data category." A discussion of the evolution of UCNI may be found in National Research Council. 1995. *A Review of the Department of Energy Classification Policy and Practice* (Washington, D.C.: National Academy Press).

[9] With the end of nuclear testing by the United States and the creation of the Stockpile Stewardship Program (SSP) to maintain the safety and reliability of the U.S. nuclear arsenal, knowledge from a number of other unclassified areas of scientific research, such as advanced computing, became essential for the large-scale simulations and other measures that were part of the SSP. For scientists in those areas, work in the national laboratories has caused significant tensions between security measures and maintaining contact with the unclassified research community from which advances relevant to their work will come. These issues are discussed in Center for Strategic and International Studies. 2002. *Science and Security in the 21st Century: A Report to the Secretary of Energy on the Department of Energy Laboratories* (Washington, D.C.: The CSIS Press).

[10] For a discussion of the barriers to nuclear weapons proliferation, see MacKenzie, D. and G. Spinardi. 1995. "Tacit knowledge, weapons design, and the uninvention of nuclear weapons." *American Journal of Sociology* 101 (July):44-99.

[11] The topic is discussed briefly in National Research Council, Committee on Science and Technology for Countering Terrorism. 2002. *Making the Nation Safer: The Role of Science and Technology in Countering Terrorism* (Washington, D.C.: The National Academies Press), Chapter 2. It is worth noting that public discussion of these possibilities is problematic in the same way that the publication of the mousepox and polio virus research has been.

[12] Dam, K., and H. Lin, eds. 1996. *Cryptography's Role in Securing the Information Society* [CRISIS] (Washington, D.C.: National Academy Press), especially Ch. 5. Following the September 11 attacks, the idea of requiring the "clipper chip" was briefly resurrected in Congress, but then dropped.

[13] CRISIS, p. 417. See also *Scientific Communication and National Security*, Appendix E (fn. 8).

[14] There is, however, a considerable distance between the mathematics of devising an encryption algorithm and constructing a secure working system; in most systems there are multiple vulnerabilities outside the encryption process. For example, users may share their passwords or keep them on a note posted on their computers.

[15] U.S. nuclear nonproliferation policy has tried to discourage the development of civilian nuclear power in states it considers would-be proliferators, however, in part to deny such states "cover" for their programs and in part because the technical skills and knowledge obtained through working on civilian nuclear power are considered useful in a general sense for those trying to develop nuclear weapons.

[16] Monatersky, R. 2002. "Publish and Perish? As the Nation Fights Terrorism, Scientists Weigh the Risks of Releasing Sensitive Information," *The Chronicle of Higher Education*, October 11.

[17] "Openness in an insecure world." 2003. *The Lancet Infectious Diseases* 3 (February). See also discussion in Chapter 1 of this report, p. 25.

[18] The financial costs could be considerable. An estimate of the costs of the U.S. nuclear weapons program between 1940 and 1995 suggests a rough figure of $75 billion for secrecy. Representative items on the list that would apply to the control of biological select agents include costs for screening personnel; for secure filing cabinets; for guards; and for routine inventories of controlled material. Schwartz, S.I. 1995. "Four trillion dollars and counting," *Bulletin of the Atomic Scientists* 51 (Nov/Dec):32-52, especially p. 50-51.

[19] For example, USAMRIID has only 650 employees, a third of which are Army officers. Enserink, M. 2002. "On Biowarfare's Frontline," *Science* 296 (5575): 1954-1956. For a sense of the activities supported at Fort Detrick, see Covert, N. 1993. *Cutting Edge: A History of Fort Detrick, Maryland, 1943-1993*, Public Affairs Office, Headquarters U.S. Army Garrison, Fort Detrick, MD. In addition to the work performed in defense laboratories, the Army and other services and agencies fund research on biodefense in industry and universities.

[20] Association of American Universities. March 2003. Definitions and Regulations Involved in the Classified-Sensitive Information-Unclassified Debate.

[21] Meridian Corporation. 1992. "Classification Policy Study," Report prepared for the U.S. Department of Energy Office of Classification, Washington, D.C.; July 4: p. 87.

[22] The Invention Secrecy Act of 1951, 35 U.S.C. 181-188. Available at http://www.fas.org/sgp/othergov/invention/35usc17.html.

[23] The Project on Government Secrecy of the Federation of American Scientists. Available at http://www.fas.org/sgp/othergov/invention.

[24] See Aftergood, S. "Government Secrecy and Knowledge Production: A Survey of Some General Issues," in J. Reppy, ed., *Secrecy and Knowledge Production*, Peace Studies Occasional Paper # 23, Cornell University Peace Studies Program, October 1999.

[25] See footnote 2.

[26] "Fundamental" research is defined as "basic and applied research in science and engineering, the results of which ordinarily are published and shared broadly within the scientific community, as distinguished from proprietary research and from industrial development, design, production and product utilization, the re-

sults of which ordinarily are restricted for proprietary or national security reasons." See National Security Decision Directive 189, September 21, 1985. Available at http://www.fas.org/irp/offdocs/nsdd/nsdd-189.htm.

[27] Rice, C. Assistant to the President for National Security Affairs, Letter to Dr. Harold Brown, November 1, 2001. Available at http://www.fas.org/sgp/bush/cr110101.html. John Marburger, Director of the Office of Science and Technology Policy, Executive Office of the President, reaffirmed NSDD-189 in a speech to a workshop on "Scientific Openness and National Security" at The National Academies on January 9, 2003. Available at http://www.ostp.gov/html/new.html.

[28] The CSIS Commission on Science and Security in the 21st Century identified at least 20 types of information that could be considered "sensitive" within the Department of Energy, most without consistent, department-wide definitions or application. See Center for Strategic and International Studies, op. cit., p. 55.

[29] Knezo, op. cit., p.10.

[30] U.S. Congress. 1996. The Freedom of Information Act, 5 U.S.C. § 552, as Amended by P.L. 104-231, 110 Stat. 2422, October 2, which states that "Geological and geophysical information and data, including maps, concerning wells" may be withheld under FOIA.

[31] See report of the Ad Hoc Faculty Committee on Access to and Disclosure of Scientific Information. 2002. *In the Public Interest*. Massachusetts Institute of Technology, June 12. Available at http://web.mit.edu/faculty/reports/publicinterest.pdf. See also Statement on "Science and Security in an Age of Terrorism" by the presidents of The National Academies regarding "sensitive but unclassified" information, October 18, 2002. Excerpt cited in Chapter 4. Available at http://www4. nationalacademies.org/news.nsf/isbns10182002b?Open Document.

[32] The Public Health Security and Bioterrorism Preparedness and Response Act of 2002 calls for a national registry of facilities holding select agents, and requires that information about the sites be kept secret. Ronald M. Atlas, immediate past president of the American Society for Microbiology, points out that this runs counter to the requirement that Institutional Biosafety Committees operate with community participation and maximum transparency. Atlas, R.M. "Applicability of the National Institutes of Health Recombinant DNA Advisory Committee paradigm for reducing the threat of bioterrorism," Draft paper prepared for the Controlling Dangerous Pathogens Project, CISSM, School of Public Affairs, University of Maryland, p. 26.

[33] Card, A.H. Jr. 2002. "Action to safeguard information regarding weapons of mass destruction and other sensitive documents related to Homeland Security," March 19. Memorandum for the heads of executive departments and agencies. Available at http://www.fas.org/sgp/bush/wh031902.html. See also Matthews, W. 2002."OMB weighs info classification," *Federal Computer Week* September 16. Guidance accompanying the memo states that: "The need to protect such sensitive information from inappropriate disclosure should be carefully considered on a case-by-case basis, together with the benefits that result from the open and efficient exchange of scientific, technical, and like information."

[34] U.S. Congress. Homeland Security Act of 2002. P.L. 107-296 (November 25). Available at http://www.cio.gov/documents/pl_107_296_nov_25_2002.pdf.

[35] An excellent review of the legislation, policies, and issues surrounding sensitive information may be found in the Congressional Research Service study by G. Knezo, op. cit.

[36] *Washington Post*. February 13, 2000.The potential penalties included civil fines of up to $100,000. "Sloppy Secrecy,"p. B6.

[37] U.S. Department of Energy, Office of Security Affairs. 1995. "Safeguards and Security Glossary of Terms," December 18 (cited in CSIS, op. cit., p. 56). Available at http://www.directives.doe.gov/pdfs/nnglossary/termss_z.pdf.

[38] Department of Energy. 1999. "Sensitive Countries and Sensitive Subjects List," Memorandum for heads of departmental elements and contractor organizations, July 27, p. 7-9.

[39]Laplante, P.R. 2003. The DOE OUO Program, Briefing to the Roundtable on Scientific Communication and National Security, The National Academies and the Center for Strategic and International Studies. Washington, D.C., June 19.

[40] Abraham, S. 2003. "Memorandum for heads of all departmental elements," May 12. Available at http://www.fas.org/sgp/othergov/doe/sec051203.pdf. In his memo Secretary Abraham cited the recommendations of the CSIS commission cited in footnote 9.

[41] "Generally, technologies subject to the Export Administration Regulations (EAR) are those which are in the United States or of U.S. origin, in whole or in part. Most are proprietary. Technologies which tend to require licensing for transfer to foreign nationals are also dual use (i.e., have both civil and military applications) and are subject to one or more control regimes, such as national security, nuclear proliferation, missile technology, or chemical and biological warfare. See "Deemed Exports Questions and Answers," Bureau of Industry and Security, Department of Commerce. Available at http://www.bxa.doc.gov/DeemedExports/DeemedExportsFAQs.html#TopofPage.

The International Traffic in Arms Regulations (ITAR), administered by the Department of State, controls the export of technology, including technical information, related to items on the U.S. Munitions List. Unlike the EAR, however, "publicly available scientific and technical information and academic exchanges and information presented at scientific meetings are not treated as controlled technical data." See Knezo, op. cit., p. 4.

[42] *Ibid.*

[43] Zilinskas, R. and J.B. Tucker. 2002. "Limiting the contribution of the open scientific literature to the biological weapons threat." *Online Journal of Homeland Security*, December. Available at http://www.homelandsecurity.org/journal/Articles/tucker.html.

[44] Jackson, R.J., A.J. Ramsay, C.D. Christensen, S. Beaton, D.F. Hall, and I.A. Ramshaw. 2001. "Expression of Mouse Interleukin-4 by a recombinant *Ectromelia* virus suppresses cytolytic lymphocyte responses and overcomes genetic resistance to Mousepox." *Journal of Virology* 75:1205-1210.

[45] Schemo, D.S. 2002. "Sept. 11 Strikes at Labs' Door," *New York Times*, August 13, p. F1, 2. See also Shea, D.A. 2003. "Balancing Scientific Publication and National Security Concerns: Issues for Congress." Congressional Research Service, Report RL31695, January 10. In a hearing before the House Committee on Science on October 10, 2002, the President's Science Adviser, Dr. John Marburger, stated: "I'm

aware that there is an impression that the administration is considering a policy of pre-publication review of sensitive federally funded research. This is incorrect— this is not the thrust of the considerations, and it's important to note that this process is in the formative stage." See "President's science advisor clarifies plan for sensitive research." Available at http://www.house.gov/science/press/107/107-299.htm.

[46] An example of this dilemma is illustrated by the controversy over a report on the risks of agricultural bioterrorism completed by the National Academies in late 2002. The report is National Research Council. 2002. *Countering Agricultural Bioterrorism.* (Washington, D.C.: The National Academies Press). For an account of the controversy, see Monastersky article cited in footnote 16.

[47] The ASM policy is not restricted to select agents exclusively. At the present time, all manuscripts addressing research conducted on select agents are flagged but others may be as well. At the NAS-CSIS international workshop on "Scientific Openness and National Security" (January 9, 2003 in Washington, D.C.), Donald Kennedy, editor of *Science*, indicated that his journal had a system of review that used outside consultants.

[48] Atlas, R.M. Email communication. August 15, 2003.

[49] The National Academies and Center for Strategic and International Studies. 2003. "Scientific Openness and National Security," January 9. Further information about the workshop, including transcripts of presentations, is available at http://www7.nationalacademies.org/pga/Scientific_Openness_Homepage.html.

[50] Journal Editors and Authors Group. 2003. "Statement on Scientific Publication and Security," *Science Online* 299 (5610):1149. Available at http://www.sciencemag.org/cgi/reprint/299/5610/1149.pdf. This statement also appeared in the February 18, 2003 issue of the *Proceedings of the National Academy of Sciences* and the February 20, 2003 issue of *Nature.*

[51] See, for example, Hesman, T. 2003. "Critics question journals' bow to security," *St. Louis Post-Dispatch*, February 23; Kennedy, D. 2003. "To Publish or Perish?" *The Times Higher Education Supplement*, February 21.

[52] Ad Hoc Faculty Committee on Access to and Disclosure of Scientific Information. 2002. *In the Public Interest.* Massachusetts Institute of Technology, June 12. Report is available at http://web.mit.edu/faculty/reports/publicinterest.pdf.

4

Conclusions and Recommendations

INTRODUCTION

The preceding chapters have reviewed the nature of the threat associated with "dual use" knowledge in the life sciences, the current regulatory environment for the conduct and reporting of genetic engineering research in the life sciences, both domestically and internationally, and various information control regimes developed over the last 60 years in the United States. The Committee has concluded that existing domestic and international guidelines and regulations for the conduct of basic or applied genetic engineering research may ensure the physical safety of laboratory workers and the surrounding environment from contact with, or exposure to, pathogenic agents or "novel" organisms. However, they do not currently address the potential for misuse of the tools, technology, or knowledge base of this research enterprise for offensive military or terrorist purposes. In addition, no national or international review body currently has the legal authority or self-governance responsibility to evaluate a proposed research activity prior to its conduct to determine whether the risks associated with the proposed research, and its potential for misuse, outweigh its potential benefits.

After extensive deliberation, the Committee recognized the importance of educating the biotechnology research community about the potential dangers posed by dual use of new technologies. Rather than considering methods to identify and prohibit certain areas of research, we believe the community should work together with government agencies to develop communication channels so that both are aware of potential problems. The Committee has concluded that a system is needed to build these channels of communication and to provide greater oversight for the research enterprise. The significant increases in funding that will be going to research on biodefense—precisely the sort of research likely to pose the most severe dual use dilemmas—reinforce the argument for creating such a system. As the case studies discussed in Chapter 1 demonstrate, even experiments that have the greatest potential for diversion to offensive applications or terrorist purposes may also have potentially beneficial uses for public health promotion and defense. To proscribe such experiments without a thorough assessment of their potential risks and benefits carries the possibility for hindering our ability to detect, identify, and defend against the new threat environment.

The system we are proposing would establish a number of stages at which experiments and eventually their results could be reviewed to provide reassurance that advances in biotechnology with potential applications for bioterrorism or biological weapons receive responsible oversight. The system relies heavily upon voluntary self-governance by the scientific community and expansion of an existing regulatory process that itself grew out of an earlier response by the scientific community to the perceived risks associated with gene-splicing research.

The heart of the system would be a set of guidelines to help identify research that could raise concerns because of its potential for diversion to offensive military applications. The concept behind these guidelines is that they will provide criteria that can assist knowledgeable scientists, editorial boards of scientific journals, and funding agencies in weighing the potential for offensive applications against the expected benefits of an experiment in this arena. It is important to realize, however, that identifying these concerns will not prevent a determined nonstate actor or individual from doing harm. Moreover, the Committee was adamant that these concerns should not be interpreted as defining a category of "sensitive but unclassified" research. Rather, like the NIH Guidelines for Research Involving rDNA Molecules established in the 1970s to guide research in a then-new and possibly risky technology, they can serve as the basis for a continuing dialogue between the members of the scientific community, the national security community, and the public. And, like the rDNA Guidelines, they must be applied on a case-by-case basis, with the opportunity for revision as new knowledge and experience in their operation accumulate.

KEY ASSUMPTIONS

In developing the system outlined in this chapter, the Committee based its recommendations on several key assumptions. Each is discussed in turn.

Many more experiments will explore the virulence factors of bacteria and viruses. The great majority of experiments on pathogenic bacteria or viruses are performed to ascertain exactly what makes the microbes pathogenic and virulent. Scientists are thus continuously exploring the ways that turning certain genes "on" and "off" enable these agents to be transmissible or cause disease in an appropriate host organism. Moreover, the concern over bioterrorism has stimulated the government to provide significantly increased funding to help combat infectious disease. The Fiscal Year 2003 budget passed by Congress and signed into law by President Bush in January 2003 added $6 billion–$10 billion, spread across a number of agencies, to biodefense research in the United States.[1] The NIH, for example, received $1.5 billion for biodefense research. Internationally, other countries are also increasing their investments in civilian bioterrorism defense research. These increased domestic and international investments in basic and applied public health and bioterrorism defense research will inevitably create an increased number of research activities that raise concerns about misuse. This increased activity will also undoubtedly increase the number of research practitioners in this ever-expanding field of investigation with a corresponding increase in the number of articles appearing in the peer-reviewed literature.

Scientific evaluation of the risks is essential. In Chapter 1 we described and provided brief assessments of the dual use dilemmas presented by three recently published experiments. Although quite different, each of these cases has generated significant controversy. In the judgment of many scientists, the publication of the synthetic poliovirus paper presented no great contribution to the field of virology and no obvious advantages to a bioterrorist.[2] The risks for potential misuse generated by the information in the other two papers are certainly greater. This is not to say they should not have been published. Rather, the Committee believes these cases illustrate that, to balance these risks against the obvious benefits, one must depend upon expert scientific judgment. In fact, the paper describing the engineering of the mousepox was judged both by local scientific officials in Australia and by the editorial board of the *Journal of Virology* to have scientific merit and, on balance, to provide important information required for progress in fighting disease.[3] The third paper stimulated an accompanying commentary by the journal when it was published. The commentary concluded that the benefits of the original research contribution to the understanding of the complement system far outweighed the risks that the information could be "misused."[4] But making such judgments requires scientific training and knowledge—and expertise in one field may not always provide sufficient understanding of the relevance of research results in another. The qualitative and case-by-case nature of these judgments is a primary reason the committee believes it is better to rely on self-governance to manage this aspect of the problem rather than to attempt to define appropriate or inappropriate research via regulation. As discussed above, key aspects of the system we propose are intended to augment the existing statutory and regulatory framework for controlling biological materials and personnel through voluntary arrangements addressing research issues.

Only an international set of standards will help to minimize the misuse of biotechnology. Although the focus of this report is on the United States, this country is only one of many pursuing biotechnology research at the highest level. The techniques, reagents, and information that could be used for offensive applications are readily available and accessible. And the expertise and know how to use or misuse them is distributed across the globe. Without international consensus and consistent guidelines for overseeing research in advanced biotechnology, limitations on certain types of research in the United States would only impede the progress of biomedical research here and undermine our own national interests. It is entirely appropriate for the United States to develop a system to provide oversight of research activities domestically, but the effort will ultimately afford little protection if it is not adopted internationally. This is a challenge for governments, international organizations, and the entire international scientific community. Efforts to meet that challenge are under way, but they must be quickly expanded, strengthened, and harmonized.

RECOMMENDATIONS

The system we propose for the United States consists of a number of filters for research proposals and publication of results that would cumulatively serve to protect against potential misuse yet enable important research activities to go forward.[5] The key initial filter is awareness in the research community of categories of research that should raise concerns and collective community commitment to actively manage such research. Voluntary restraint based on awareness should be supplemented by review through existing bodies, namely an Institutional Biological Safety Committee/Recombinant DNA Advisory Committee process augmented to include the assessment of the potential for misuse as a criterion for approval or denial of proposed experiments. At the stage of publication, we recommend enhancing and expanding the process begun by the editors of a number of the leading scientific journals in February 2003. Finally, since this new system relies on both regulatory and voluntary elements, and involves issues and relationships with which the life sciences community has little experience compared to its colleagues in other fields, we recommend creation of a National Science Advisory Board for Biodefense (NSABB) to provide advice and assessments to the government and the scientific community as the system of review we are proposing develops. We recognize that successfully implementing the system we propose will require significant additional resources at each stage; we do not attempt to provide an estimate of those costs. Otherwise, concerns for unfunded mandates could be a significant barrier to full consideration of the proposals by the scientific community.

In making its recommendations, the Committee has sought to strike a balance and propose processes and mechanisms that will raise awareness and alarms when needed, without unduly constraining the practice, processes, and products of the life sciences research enterprise. We believe that such a system in the United States could also serve as a model for similar restraint in other countries.

Recommendation 1: Educating the Scientific Community
We recommend that national and international professional societies and related organizations and institutions create programs to educate scientists about the nature of the dual use dilemma in biotechnology and their responsibilities to mitigate its risks.

Adequately addressing the potential risks that research in advanced biotechnology could be misused by hostile parties will require educating the community of life scientists, both about the nature of these risks and about the responsibilities of scientists to address and manage them. At present, awareness of the potential for misuse of biological knowledge varies widely in the research community. Researchers currently working with select agents are already taking steps to contain these agents physically and protect against planned or unplanned harm. But most life scientists have had little direct experience with the issues of biological weapons and bioterrorism since the advent of the Biological Weapons Convention in the early 1970s, so these researchers lack the experience and historical precedent of considering the potential for misuse of their discoveries.

Fortunately, an extensive national and international network of professional societies provides the natural basis for increasing knowledge and awareness about the potential risks of research in advanced biotechnology. These societies hold numerous professional meetings to share the results of research and address issues of concern to the research community. We recommend that the societies undertake a regular series of meetings and symposia at these gatherings, in the United States and overseas, to provide both knowledge and opportunities for discussion. It could be useful for one of the major professional societies or science policy organizations to convene a meeting of all the major societies to discuss how best to implement such a program. Industry groups and associations of higher education and research could also usefully undertake the education of their members about the risks and their implications for research practices.

Substantive knowledge of the potential risks is not sufficient, however. The Committee believes that biological scientists have an affirmative moral duty to avoid contributing to the advancement of biowarfare or bioterrorism. Individuals are never morally obligated to do the impossible, and so scientists cannot be expected to *ensure* that knowledge they generate will never assist in advancing biowarfare or bioterrorism. However, scientists can and should take reasonable steps to minimize this possibility. The Committee believes that it is the responsibility of the research community, including scientific societies and organizations, to define what these reasonable steps entail and to provide scientists with the education, skills, and support they need to honor these steps.

These principles should be added to the codes of ethics of relevant professional societies. Most scientists are familiar with and carefully respect the moral norms of their profession that focus on the pursuit of truth and the advancement of science. Often placed under the heading of research integrity, prohibitions against fraud and plagiarism, as well as affirmative duties of openness in the sharing of findings, are well understood. Concerns about potential conflicts of interest and respect for intellectual property are similarly well appreciated, if not as clearly delineated. The addition of the moral duty we endorse here to those more familiar to the scientific community is not as novel as it may appear at first. As some scientific societies have recognized, scientists also have a general moral duty to use their knowledge and skill for the advancement of human welfare.[6] We are only providing a specification of that general moral responsibility.

We believe further that scientists have an obligation to inculcate these moral duties in the next generation, both by example and by specific education and evaluation of their trainees. Other models of training in social responsibilities need to be explored, for example from the law and from medicine. In the law, most students sit for a multistate professional responsibility exam. In medicine, many specialty boards now examine young physicians in ethics as well as in medical skill and knowledge. Scientists will need assistance in learning about these other models, but they need to take charge of how best to educate their own next generation.

Scientists also should be willing to assist efforts to integrate the advancement of knowledge with the protection of national security by volunteering their time to sit on relevant peer review committees and national bodies, much as scientists contribute to advancing science currently by serving on study sections and as reviewers for professional journals. Finally, if scientists are to embrace the moral responsibilities outlined here, their home institutions must provide accommodation and support. Service in review of protocols and in student training must not only be encouraged but also rewarded.

Recommendation 2: Review of Plans for Experiments
We recommend that the Department of Health and Human Services (DHHS) augment the already established system for review of experiments involving recombinant DNA conducted by the National Institutes of Health to create a review system for seven classes of experiments (the Experiments of Concern) involving microbial agents that raise concerns about their potential for misuse.

This part of the system includes both the criteria for deciding which experiments will be subject to review and the process by which the review will take place.

The Criteria for Review

The experience with rDNA experiments emphasizes the importance of guidelines developed by the scientific community itself. The guidelines for work with rDNA promulgated after the Asilomar Conference in 1975 have proven remarkably flexible and effective. The logical structure of such a system, keyed to the probability of harm associated with exposure(s) to genetically modified organisms—which changed as risk perceptions changed—has been integral to their success. The guidelines have prevented any untoward events and have allowed the rapid and efficient progress of the academic and commercial applications of these technologies. We now need to build upon the Asilomar experience to develop a uniform set of criteria to manage this new set of risks.

The Committee identified seven classes of experiments that it believes illustrate the types of endeavors or discoveries that will require review and discussion by informed members of the scientific and medical community before they are undertaken or, if carried out, before they are published in full detail. These categories represent experiments that are feasible with existing knowledge and technologies or with advances that the Committee could anticipate occurring in the near future. Some of them represent the types of naturally occurring genetic changes in pathogens that have led to disease pandemics such as the "Spanish Flu" in 1917-1918 or the recently recognized disease "severe acute respiratory syndrome" (SARS) but that could now be engineered in the laboratory. Others have been part of the history of biowarfare research and development. Furthermore, carrying out these types of experiments could, in some instances, lead to the potential for great damage without significantly advancing our knowledge in ways that would either greatly increase our ability to defend against them or our ability to promote human health by preventing, diagnosing, or treating common human diseases. The concerns deal with infectious agents or their products because the Committee believes that self-replicating agents or their products pose the most imminent biological threat.

The seven areas of concern listed here only address potential microbial threats. Of course, modern biological research is much broader, encompassing all of the health sciences, agriculture, and veterinary science. It also includes diverse industries such as those that manufacture pharmaceuticals, cosmetics (e.g., Botox), and soft drinks (e.g., citric acid production). Moreover, all of these areas are changing rapidly. The great diversity as well as the pace of change makes it imprudent to project the potential both for good and ill too broadly and too far into the future. Therefore, the Committee has initially limited its concerns to cover those possibilities that represent a plausible danger and has tried to avoid improbable scenarios. Over time, however, the Committee believes that it will be necessary not only to expand the experiments of concern to cover a significantly wider range of potential threats to humans, animals or crops but also to include oversight of work conducted for or performed within the private sector as well as non-NIH government facilities and sponsored activities that are not already voluntarily complying with the Guidelines.

Experiments of Concern would be those that:

1. Would demonstrate how to render a vaccine ineffective. This would apply to both human and animal vaccines. Creation of vaccine-resistant smallpox virus would fall into this class of experiments.

2. Would confer resistance to therapeutically useful antibiotics or antiviral agents. This would apply to therapeutic agents that are used to control disease agents in humans, animals or crops. Introduction of ciprofloxacin resistance in *Bacillus anthracis* would fall into this class.

3. Would enhance the virulence of a pathogen or render a nonpathogen virulent. This would apply to plant, animal, and human pathogens. Introduction of cereolysin toxin gene into *Bacillus anthracis* would fall into this class.

4. Would increase transmissibility of a pathogen. This would include enhancing transmission within or between species. Altering vector competence to enhance disease transmission would also fall into this class.

5. Would alter the host range of a pathogen. This would include making nonzoonotics into zoonotic agents. Altering the tropism of viruses would fit into this class.

6. Would enable the evasion of diagnostic/detection modalities. This could include microencapsulation to avoid antibody-based detection and/or the alteration of gene sequences to avoid detection by established molecular methods.

7. Would enable the weaponization[7] of a biological agent or toxin. This would include the environmental stabilization of pathogens. Synthesis of smallpox virus would fall into this class of experiments.

The Review Process

The NIH Guidelines require creation of an Institutional Biosafety Committee (IBC) when research is conducted at or sponsored by an entity receiving any NIH support for recombinant DNA research. Most of the 400 or so IBCs registered with NIH are at institutions that are subject to the NIH Guidelines and for whom IBC registration is mandatory. While most of these institutions are academic, some industry-based IBCs are registered with NIH as a consequence of receiving NIH support. In other instances, companies voluntarily comply with the NIH Guidelines as a means of demonstrating a commitment to a "gold standard" for safety practices. Several Federal agencies and laboratories have made compliance with the NIH Guidelines a condition of their support of intramural and extramural research projects.[8] Furthermore, a number of federal IBCs are registered with NIH.

All the experiments that fall within the seven areas of concern should currently require review by an Institutional Biosafety Committee (IBC), a process described in detail in Chapter 2. This review should take place regardless of the source of funding, whether the institution doing the research is public or private, and whether it is a university, government laboratory, or business. We thus recommend relying on the system of IBCs as the first review tier for experiments of concern. We note that funding agencies also have a potentially important role to play in flagging experiments of concern at the proposal review stage.

Like the broader life sciences community, the members of the IBCs will require substantial education in the potential risks associated with advanced biotechnology research in order to handle this task competently. Many IBCs may need to add expertise in immunology, virology, pathology, and epidemiology to undertake this new responsibility. Some of this is already occurring as part of implementing the requirement of the Bioterrorism Response Act, but more will need to be done. To ensure the most consistent application possible of the review process—and as a reassurance to the research scientists subject to the new review—regular opportunities for members of IBCs to gather and discuss the process should be provided on a continuing basis.

We recommend that the form researchers now use to submit their experimental designs to the IBC be amended to include another category where researchers would designate whether, in their judgment, their proposed projects fit into an area of concern. The IBC would then review that issue along with the other aspects of the project that it is evaluating, carefully weighing potential benefits versus potential danger. Occasionally, the IBC may discover that what is proposed is forbidden under current guidelines, and would not approve the research. In most cases, however, it would either designate the project as acceptable to move forward or as one raising concerns that need further consideration.

Experiments that need further consideration would be referred to an expanded NIH RAC and possibly the Director of the National Institutes of Health for approval or denial of permission to proceed with the proposed experiment. We recommend this route because so many of the experiments in the area of concern would fall under the purview of the RAC already and because it has an established track record of facilitating research while protecting public safety. We propose that when the RAC takes up this new duty, it initially translate our categories of experiments of concern into a set of guidelines for IBCs to use. It should then improve and update these guidelines as needed as its experience with the process grows. The RAC will need substantial new resources to take on this additional task, and both it and the IBCs may need to incorporate new expertise to handle the task.

Under our recommendation, the RAC would begin to review some projects in the areas of concern from all relevant research institutions. This would be a substantial expansion from its current jurisdiction over research funded by the NIH and those institutions that comply voluntarily. The RAC guidelines would thus need to be revised and reproposed in the Federal Register to reflect this expanded scope and mandate.

As we envision this review of the experiments of concern, when an IBC refers a project to the RAC, the RAC would carefully weigh the potential benefits and dangers of the project, and come to its own independent judgment. The RAC may approve some projects referred by IBCs to go forward at this point, recommend that the research not be undertaken, or that modifications be made to the research design to minimize the potential risks.

Recommendation 3: Review at the Publication Stage
We recommend relying on self-governance by scientists and scientific journals to review publications for their potential national security risks.

By the time a manuscript is submitted for publication, substantial information about the research may have already been disseminated through informal professional contacts, presentations of preliminary results at scientific meetings, or consultations with colleagues. This is why the Committee recommends a system that can address research at its earliest stages, and why it is so important to make scientists aware of their personal responsibilities to consider the balance of risks and benefits in research they consider undertaking. Nevertheless, publication of research results provides the vehicle for the widest dissemination, including to those who would misuse them. It is thus appropriate to consider what sort of review procedures can be put in place at the stage of publication to provide another layer of protection. The Committee believes strongly that this part of the system should be based on the voluntary self-governance of the scientific community rather than on formal regulation by government.

Proposals to limit publications have caused great concern and controversy among both scientists and publishers. The norm of open communication is one of the most powerful in science. To limit the information available in the methods section of journal articles would violate the norm that all experimental results should be open to challenge by others. But not to do so is potentially to provide important information to biowarfare programs in other countries or to terrorist groups.

Journals in the life sciences have already responded to the challenge in a variety of ways; the procedures that a number of leading publications have undertaken to screen manuscripts were discussed in Chapter 3.[9] The joint statement by editors of four major journals in the life sciences issued in February 2003 was a major step toward developing this part of the system of oversight the Committee believes will be necessary. It was also an important example of the ability of the scientific community to address the potential risks of its activities.

Ultimately, any process to review publications for their potential national security risks would have to be acceptable to the wide variety of journals in the life sciences, both in the United States and internationally. The Committee believes that continued discussion among those involved in publishing journals—and between editors and the national security community—will be essential to creating a system that is considered responsive to the risks but also credible with the research community. The national advisory board recommended in the next section could serve as a forum for such discussions and for creating greater consensus in the scientific community about the appropriate role of and process for review at the publication stage.

On the broader question of classification, the Committee believes that the principle set out by the Reagan Administration in 1985 in National Security Decision Directive 189 remains valid and should continue to be the basis for U.S. policy. As discussed in Chapter 3, the policy states that:

"to the maximum extent possible, the products of fundamental research remain unrestricted. ... where the national security requires control, the mechanism for control of information generated during federally-funded fundamental research in science, technology and engineering at colleges, universities and laboratories is classification."[10] The Committee's support for self-governance by the scientific community through appropriate reviews by journals and other publication outlets should not be construed as endorsing the creation of "sensitive but unclassified" information in the life sciences. We believe that the risks of a chilling effect on biodefense research vital to U.S. national security as the result of inevitably general and vague categories is at present significantly greater than the risks posed by inadvertent publication of potentially dangerous results. A system of review based in scientific self-governance can, we believe, effectively address the security risks without discouraging scientists from participating in important biodefense research.

Recommendation 4: Creation of a National Science Advisory Board for Biodefense

We recommend that the Department of Health and Human Services create a National Science Advisory Board for Biodefense (NSABB) to provide advice, guidance, and leadership for the system of review and oversight we are proposing.

The NSABB would serve a number of important functions for both the scientific community and the government. At the most general (strategic) level, it would serve as a point of continuing dialogue between the scientific community and the national security community and as a forum for addressing issues of interest or concern. At the operational (tactical) level, it would provide case-specific advice on the oversight of research and the communication and dissemination of life sciences research information that is relevant for national security and biodefense purposes. Because of its important bridging functions, its members should include both leading scientists and national security experts, including those with experience in managing scientific research in federal agencies. Particularly in the early phases of its work, it would be desirable to include among the Board's members a few scientists or engineers from research fields long associated with applications to national security.

In terms of the regulatory aspects of the operation of our proposed system, we recommend that the Board periodically review and suggest updates to the "Experiments of Concern." We also recommend that the Board review and suggest updates to the list of "select agents" and to policies regarding the international exchange of biological agents. A review of the select agents list by DHHS is already required every two years but the Board could serve a useful and important function by providing an independent assessment as an input to that process.

For the system's self-governing phases, we recommend that the NSABB serve as a resource. This could include aiding the professional societies in developing education programs, as well as providing a convening mechanism. It could also include assisting those producing publications in the life sciences. The Board could provide a convening mechanism for journal editors, organizing periodic discussions among them as they develop and evaluate their review processes. The Board could review and comment on proposed procedures on request, and perhaps serve as a clearinghouse so that journals that have not already adopted review procedures could have ready access to examples of what others are doing. It would be very important for the Board to reach out beyond the United States to the many international publications in the life sciences and to find ways to include their leaders in discussions. The Board might also provide advice on request about particular manuscripts that raise concern, perhaps by organizing small groups of experts to assess the trade-offs between the scientific merits of the research, especially that with the potential to advance knowledge relevant to biodefense, and the risks of publishing information that might assist terrorists or proliferatant states.

So far, we have only discussed the functions of the Board that relate to the potential risks of research in advanced biotechnology. But we also recommend that the Board have the capacity to advise the government on how the life sciences can contribute to alleviating the risks of bioterrorism and biological weapons through new research in areas such as vaccine and antibiotic development, new detection devices and technologies, and preventative public health measures. This advisory function would serve as a continuous reminder that any system of review and oversight must operate in ways that do not put the United States—and the world—at risk of losing the great potential benefits of biotechnology. Having a Board that was informed and aware of the latest research developments, even including manuscripts not yet published, would provide the capacity for "early warning," alerting the government to the risks of new findings or techniques that should be met by focusing research resources on appropriate responses or countermeasures.

We considered a number of options for the organizational location for the NSABB. There are clear trade-offs between an independent board that offers its advice to government and one that is a formal advisory body to one or more federal agencies. The relationship between the life sciences community and the national security community is new and still tenuous, with significant potential for suspicion and misunderstanding on both sides. The topics the Board will address are both sensitive and controversial within the scientific community, and there is a risk that a formal federal role could raise concerns about the NSABB's capacity to offer genuinely independent advice. Whatever the home, the organization that houses the Board must have high credibility with the scientific community, since its engagement is essential to success. But it must also be able to command the cooperation and trust of the national security community and of the full range of U.S. research facilities, public and private, and publications. A formal attachment to the U.S. government would ensure access to the relevant high-level decision makers. The host organization should also have sufficient international standing to gain the necessary cooperation from the research communities of other countries. Another important consideration is the suitability of the organization for conducting some closed deliberations, while overall maintaining transparency and public trust in the process.

No solution meets all the criteria, but on balance we believe that the logical organizational location for the NSABB is within the Department of Health and Human Services providing advice to the secretary of that Department.[11] DHHS already has a leading role in biotechnology research, particularly that related to the Experiments of Concern. Location within the DHHS would also connect the Board directly to the other parts of our proposed system, the RAC and the IBCs, while not limiting its capacity to work with other relevant agencies or private groups. In addition, this approach would fit within the division of labor created under the Bioterrorism Response Act, where HHS provides tactical advice—as the NSABB would do on specific issues and cases—and the Department of Homeland Security is charged with formulating overall strategy. We note that the Board will require significant financial resources to carry out its responsibilities, although the Committee did not attempt to estimate an amount.[12]

It would be important for the Board to monitor the development and operation of the system we recommend and perhaps of other processes that the government or private organizations may put in place as well. The substantial expansion of funding for research in biodefense now in progress and anticipated suggests that it will be vital to assess how these new resources affect the conduct of research and to be ready to make timely adjustments. The monitoring should be done with the goal of suggesting ways to improve the system's operation and efficiency. But it should also include the possibility of proposing that parts of the system be overhauled or even eliminated if they prove ineffective or an impediment to important scientific research.

As discussed further in Recommendation 7, international coordination and cooperation will be necessary to make any effort to mitigate the risks of bioterrorism effective. Therefore, the Committee believes that the establishment of an NSABB within the United States can serve as the basis for international dialogue aimed at reducing the risks of subversion of legitimate life sciences research efforts. Review systems, comparable to the one proposed involving the IBC and RAC, already exist in many nations. These were established as an outgrowth of the Asilomar conference in 1975. In the same manner, other countries should be encouraged to establish counterparts to the NSABB so that the community of life scientists globally can work together to reduce the risks of the offensive applications of life sciences research.

Recommendation 5: Additional Elements for Protection Against Misuse
We recommend that the federal government rely on the implementation of current legislation and regulation, with periodic review by the NSABB, to provide protection of biological materials and supervision of personnel working with these materials.

The major focus of the Committee's work has been reviewing the adequacy of the current U.S. regulatory system to deal with the increased concerns about misuse of research in advanced biotechnology and recommending a system that could better address those risks. But there are other elements of the current regulatory system that the Committee believes should be reviewed and evaluated because of their important impact on the conduct of research.

Physical Containment. Absolute containment of organisms with potential for bioterrorism is not a realistic expectation. Many of these agents can be cultured directly from nature or obtained from small animals available at any pet store or exotic animal "swap meet";[13] no genetic modification is required to convert them into weapons. It is, therefore, not feasible—with the possible exception of smallpox—to prevent knowledgeable individuals from obtaining any of the agents listed on the CDC select agent list by simply increasing the physical security of the laboratory environment.

There may, however, be individuals or rogue groups who lack the expertise either to isolate or grow pathogenic organisms, suggesting that cost-effective efforts should be made to limit access to them. Safeguarding the collections of existing agents is an obvious priority that in large measure has been addressed through recently passed legislation and implementing regulations. The CDC's and APHIS's designation of certain pathogens as "select agents" is an appropriate starting point for identifying strains and isolates that need to be secured. Additional agents, some of which have only recently been isolated, could be added to the list; agents might also be removed from the list if their potential for misuse is no longer considered a serious risk. Appropriate regulations should be enforced through the existing institutional biosafety committees.

It is crucial to avoid well-meaning but counterproductive regulations on pathogens. For example, regulations that force or provide a strong incentive for scientists to purge archival stocks of human, plant, and animal pathogens may deprive of us material that could be critical in the forensic identification of intentionally introduced pathogens into our environment and mounting an effective defense against bioweapons agents. A similar caution exists in assessing the risks of handling DNA fragments from select agents, which may pose no potential risk at all. Rules for containment and registration of potentially dangerous materials must be based on scientific risk assessment and informed by a realistic appraisal of their scientific implications. Moreover, scientific input is essential to ensure that these rules are clear as well as responsive to periodic assessment of the current technologies and capacities.

We recommended above that the NSABB review the security designations of biological agents. The Board could also be available to provide advice on short notice about revising regulations in response to new developments. Rules governing transfer of materials between laboratories to prevent unauthorized distribution or diversion might also be regularly reviewed by the NSABB so that new threats could be recognized and responded to and unnecessary impediments identified for removal.

Trained Personnel. In some areas of technology, the limiting ingredient is the existence of trained personnel. There are two aspects to the technical expertise: general microbiological know-how and knowledge about how to weaponize bacteria or viruses. General microbiological training sufficient for culturing and growing pathogenic microorganisms at levels of significant concern is available in high school and first-year college biology courses; majors in microbiology would be sophisticated enough to grow many select organisms. It should be remembered that the procedures used to grow pathogenic bacteria are identical to those used for harmless bacteria, differing primarily in the need for precautions to ensure the safety of the workers. Moreover, training in basic microbiology is widely available outside the United States. Efforts to identify or control knowledgeable personnel within the United States are, therefore, impractical, and surveillance of such personnel would not, in our opinion, offer much security.

The procedures for admitting foreign students and scientists to the United States for study and collaborative research must reflect the importance of keeping universities as open educational environments. This must be weighed against national security concerns for limiting the spread of information to adversarial groups and admitting individuals who pose risks to domestic security. Establishing procedures for limiting visas is the role of the Department of Homeland Security and allied government agencies; universities and research scientists may provide information, but they have no investigative powers. It should be borne in mind that scientists and students, in particular those from developing countries, are likely to have a major interest in infectious diseases, because such diseases impose a devastating health burden in their home countries. Also, having colleagues and well-trained health workers in other countries increases U.S. security by enabling early access to information about emerging infections.

In June 2003 the presidents of The National Academies responded to the growing concerns that new security measures directed against foreign students, workers, and scholars could cause potentially serious damage to the conduct of science in the United States by issuing a statement that read in part:

> To make our nation safer, it is extremely important that our visa policy not only keep out foreigners who intend to do us harm, but also facilitate the acceptance of those who bring us considerable benefit. The professional visits of foreign scientists and engineers and the training of highly qualified foreign students are important for maintaining the vitality and quality of the U.S. research enterprise. This research, in turn, underlies national security and the health and welfare of both our economy and society. But recent efforts by our government to constrain the flow of international visitors in the name of national security are having serious unintended consequences for American science, engineering, and medicine. The evidence we have collected from the U.S. scientific community reveals that ongoing research collaborations have been hampered; that outstanding young scientists, engineers, and health researchers have been prevented from or delayed in entering this country; that important international conferences have been canceled or negatively impacted; and that such conferences will be moved out of the United States in the future if the situation is not corrected. Prompt action is needed.[14]

Recommendation 6: A Role for the Life Sciences in Efforts to Prevent Bioterrorism and Biowarfare.
We recommend that the national security and law enforcement communities develop new channels of sustained communication with the life sciences community about how to mitigate the risks of bioterrorism.

By signing and ratifying the Biological and Toxin Weapons Convention, the United States renounced the use and possession of such offensive weapons and methods to disseminate and deliver them. Given the increased investments in biodefense research in the United States, it is imperative that the United States conduct its legitimate defensive activities in an open and transparent manner. This should clear the way for all biomedical scientists to contribute to the development of defensive measures that would mitigate the impact of the use of such weapons against people, plants, and animals. For the scientific community to be a willing partner in biodefense research, there must be trust and understanding between the scientific community and the defense, intelligence, and law enforcement branches of government.

The recent experience with anthrax dispersal in the United States made clear that there are individuals or groups in the world who will use the most horrific weapons, including pathogenic organisms, to kill innocent people for vague and unstated political goals. Added to the already existing concern about nonstate actors seeking BW capabilities, this has put bioterrorism along with biological warfare on the front burner for both the military and civilian populations. It has also meant that groups of people who had little history of working together, such as basic biomedical scientists and the FBI and CIA, must now find a way of sharing information and expertise. The nuclear physics/Department of Defense community, which grew from a relatively small group during World War II, has had a long history of participation with intelligence and defense. Biomedical science, as already discussed, has had a different history. The intelligence and law enforcement agencies need the academic scientists both for the expertise they might provide about the nature of current agents and the potential for new ones and for the best advice on limiting the spread of new technologies that would make countermeasures even more difficult. It might be desirable for components of the national security and law enforcement communities to establish advisory boards of basic scientists and clinicians with expertise in specializations such as viral disease, bacterial pathogens, biotechnology, immunology, toxins, and public health, as well as others in the area of basic molecular biology. These advisory boards could help members of these communities keep current in relevant areas of science and technology and provide trusted sets of advisors to answer technical questions.

Recommendation 7: Harmonized International Oversight
We recommend that the international policymaking and scientific communities create an International Forum on Biosecurity to develop and promote harmonized national, regional, and international measures that will provide a counterpart to the system we recommend for the United States.

Any serious attempt to reduce the risks associated with biotechnology must ultimately be international in scope, because the technologies that could be misused are available and being developed throughout the globe. A number of countries and regional and international organizations are already moving forward to develop programs and policies on aspects of the problem; the initiatives include consultations among the parties to the BWC on best practices for the security and oversight of pathogens and toxins.[15] These approaches must be harmonized and widely adopted in order for them to be effective. Just as the scientific community in the United States must become deeply and directly engaged, the commitment of the international scientific community to these issues is needed to implement the recommendations contained in this report. Diverse groups of scientists, academicians, and policymakers must be brought together for a sustained dialogue in order to develop consensus and devise a path forward.

We do not expect our recommendations to provide a "roadmap" that could simply be adopted internationally without significant modifications or adaptations to local or regional conditions. But any effective system should include all the issues addressed by our recommendations. The Committee therefore recommends, as a next step, convening an "International Forum on Biological Security" to begin a dialogue within and between the life sciences and the policymaking communities internationally. Among the topics for this international forum are:

- Education of the scientific community globally, including curricula, professional symposia, and training programs to raise awareness of potential threats and modalities for reducing risks as well as to highlight ethical issues associated with the conduct of biological science.
- Design of mechanisms for international jurisdiction that would foster cooperation in identifying and apprehending individuals who commit acts of bioterrorism.
- Development of an internationally harmonized regime for control of pathogens within and between laboratories and facilities.
- Development of systems of review to provide oversight of research, including defining an international norm for identifying and managing "experiments of concern."
- Development of an international norm for the dissemination of "sensitive" information in the life sciences.

The Committee believes that, to be most effective, this and other forums should be sponsored by international organizations with the standing and credibility within both the policymaking and scientific communities. Different topics within this broad agenda may be more appropriate for different organizations. Potential sponsors could include the World Health Organization and the United Nations Educational, Scientific and Cultural Organization (UNESCO) as formal international governmental organizations with direct links to government policymakers. Among non-governmental scientific organizations are the International Council for Science (ICSU), created in 1931 "to identify and address major issues of importance to science and society, by mobilizing the resources and knowledge of the international scientific community;" and more recently created organizations of the world's academies of science such as the InterAcademy Panel on International Issues (IAP) and the InterAcademy Council (IAC) that seek to bring the prestige and convening capacity of these bodies to bear on crucial international problems.[16]

Finally, the Committee notes the uncertain international foundation for authoritatively addressing the issues that we have considered. No international agreement addresses the potential threats posed by the misuse of research in the biological sciences, and no intergovernmental organization has relevant oversight authority to promulgate guidelines or procedures. The Committee believes that convening an international forum to address these gaps demands international and interdisciplinary mobilization of resources and capabilities.

CONCLUSION

This report reflects the increasing attention being paid by scientists and policymakers to the potential for misuse of biotechnology by hostile individuals or nations and to the policy proposals that could be applied to minimize or mitigate those threats. The term "misuse of biotechnology" is a phrase that captures a wide spectrum of potentially dangerous activities from spreading common pathogens (e.g., spraying *Salmonella* on salad bars) to sci-fi plots of transforming pathogens into the next "Andromeda strain." Our Committee addressed one important part of this spectrum of risks: the capacity for advanced biological research activities to cause disruption or harm, potentially on a catastrophic scale. Broadly stated, that capacity consists of two elements: (1) the risk that dangerous agents that are the subject of research will be stolen or diverted for malevolent purposes; and (2) the risk that the research results, knowledge, or techniques could facilitate the creation of "novel" pathogens with unique properties or create entirely new classes of threat agents.

Throughout the Committee's deliberations there was a concern that policies to counter biological threats should not be so broad as to impinge upon the ability of the life sciences community to continue its role of contributing to the betterment of life and improving defenses against biological threats. Caution must be exercised in adopting policy measures to respond to this threat so that the intended ends will be achieved without creating "unintended consequences." On the other hand, the potential threat from the misuse of current and future biological research is a challenge to which policymakers and the scientific community must respond. The system proposed in this report is intended as a first step in what will be a long and continuously evolving process to maintain an optimal balance of risks and rewards. The Committee believes that building upon processes that are already known and trusted and relying on the capacity of life scientists to develop appropriate mechanisms for self-governance, while greatly expanding the consultation and dialogue between the science and national security communities, offers the greatest potential to find the right balance. This system may provide a model for the development of policies in other countries. Only a system of international guidelines and review will ultimately minimize the potential for the misuse of biotechnology.

NOTES

[1] Choffnes, E.R. 2002."Bioweapons: New Labs, More Terror," *Bulletin of the Atomic Scientists*, September/October:29-34.

[2] Block, S.M. 2002. "A Not-so-cheap stunt," *Science* 297 (5582):769.

[3] It should be noted that the American Society for Microbiology, which publishes the *Journal of Virology*, has as a policy to review papers not just for scientific content, but also whether, in the opinion of the Editor-in-Chief and the Publications Board, the manuscript under review describes misuses of microbiology or of information derived from microbiology. If it does, they can decline the manuscript and return it to the author.

[4] Lachmann, P. 2002: A Commentary, *Proceedings of the National Academy of Sciences* 99 (13):8461-8462.

[5] A growing number of individuals and organizations are engaged in developing proposals to regulate pathogenic materials and to provide oversight of biotechnology research activities and information dissemination at the point of publication, both in the United States and internationally. The Committee received briefings from many of them and benefited from their ideas in the course of its deliberations (see Appendix C). Among them are Jonathan Tucker, Raymond Zilinskas, George Poste, John Steinbruner, Tara O'Toole, Steve Block, Gerald Epstein, Malcolm Dando, and Mark Wheelis, and organizations such as Johns Hopkins Center for Civilian Biodefense Studies, the Center for International Security Studies at Maryland, the Center for Nonproliferation Studies of the Monterey Institute for International Studies, the Chemical and Biological Arms Control Institute, the Royal Society, and the International Committee of the Red Cross. See, for example, Kwik, G.; J. Fitzgerald; T.V. Inglesby; T. O'Toole (2003): "Biosecurity: Responsible Stewardship of Bioscience in an Age of Catastrophic Terrorism," http://rudolfo.ingentaselect.com/vl=14455723/cl=15/nw=1/rpsv/catchword/mal/15387135/v1n1/s5/p27; Tucker, J. (2003): "Preventing the Misuse of Pathogens: The Need for Global Biosecurity Standards," *Arms Control Today* June: 3-10 at http://www.armscontrol.org/act/2003_06/tucker_june03; Wheelis, M. and M. Dando (2002): "On the Brink: Biotechnology, Biodefense and the Fu-

ture of Weapons Control." http://fas-www. harvard.edu/~hsp/bulletin/ cbwcb58.pdf; Appeal of the International Committee of the Red Cross on Biotechnology, Weapons and Humanity, http://www.icrc. org/Web/eng/siteeng0.nsf/ iwpList515/274D020806432963C1256C3E005C4338 #a3; Moodie, M. (2003): "Reducing the Biological Weapons Threat: New Thinking, New Approaches; January 2003; http://www.cbaci.org/; Steinbruner, J.D. and E.D. Harris (2003): "Controlling Dangerous Pathogens," *Issues in Science and Technology,* Volume XIX, number 3; Spring 2003; Epstein, G. (2001): "Controlling Biological Warfare Threats: Resolving Potential Tensions among the Research Community, Industry, and the National Security Community," *Critical Reviews in Microbiology,* 27(4):321-354, 2001; Poste, G. (2002): "Biotechnology and Terrorism," Prospect Magazine, May 2002, http://www.prospect-magazine.co.uk...ask?accessible=yes&P%20%20 Article=11341; Zilinskas, R. and J. Tucker (2002): "Limiting the Contribution of the Open Scientific Literature to the Biological Weapons Threat, " *Journal of Homeland Security,* December 2002, http://homelandsecurity. org/journal; Block, S.M. (2001): "The Growing Threat of Biological Weapons," *American Scientist,* 89(1), (January-February 2001):28-37.

[6]Statement of the ASM Council Policy Committee: "The Council Policy Committee of the American Society for Microbiology affirms the long-standing position of the Society that microbiologists will work for the proper and beneficent application of science and will call to the attention of the public or the appropriate authorities misuses of microbiology or of information derived from microbiology. ASM members are obligated to discourage any use of microbiology contrary to the welfare of humankind, including the use of microbes as biological weapons. Bioterrorism violates the fundamental principles expressed in the Code of Ethics of the Society and is abhorrent to the ASM and its members."

[7]For the purposes of this category the term "weaponization" includes experiments that would facilitate the dissemination of a microbial pathogen as a respirable aerosol—the optimal means of delivering a putative biowarfare agent over a large geographic area. Experiments that could enhance the aerosol delivery of pathogens include, inter alia, new techniques for the microencapsulation of fragile microorganisms or the development of aerosol systems for the delivery of therapeutic drugs and vaccines. This category would also include the synthesis of viral pathogens.

[8] These include the U.S. Department of Agriculture and the Department of Veterans Affairs, as well as Department of Energy laboratories such as the Lawrence Livermore National Laboratory, the Los Alamos National Laboratory, and Sandia National Laboratories, and various VA medical centers and military research institutes such as the Uniformed Services University of Health Sciences; the Walter Reed Army Medical Center and the U.S. Army's Medical Research Institute for Infectious Diseases. See Chapter 2 for further details.

[9] See Box 3-3, Chapter 3.

[10] National Security Decision Directive 189. September 21, 1985. Available at http://www.fas.org/irp/offdocs/nsdd/nsdd-189.htm.

[11] One model for the organization and operation of the NSABB is patterned after the Advisory Committee on Immunization Practices (ACIP). The ACIP consists of 15 experts in fields associated with immunization who have been selected by the Sec-

retary of the U. S. Department of Health and Human Services to provide advice and guidance to the Secretary, the Assistant Secretary for Health, and the Centers for Disease Control and Prevention (CDC) on the most effective means to prevent vaccine-preventable diseases. The overall goals of the ACIP are to provide advice which will assist the Department and the nation in reducing the incidence of vaccine preventable diseases and to increase the safe usage of vaccines and related biological products. The Committee develops written recommendations for the routine administration of vaccines to the pediatric and adult populations, along with schedules regarding the appropriate periodicity, dosage, and contraindications applicable to the vaccines. ACIP is the only entity in the federal government which makes such recommendations. See further details available at http://www.cdc.gov/nip/acip/charter.htm.

[12] For the ACIP, the estimated annual cost for operating the Committee, including compensation and travel expenses for members (but excluding staff support) is $151,465. Estimate of annual person-years of staff support required is 2.2 at an estimated annual cost of $145,222. We would expect comparable funds to be appropriated each year for the creation and maintenance of the activities of the NSABB. Details available at http://www.cdc.gov/nip/acip/charter.htm.

[13] "The recent outbreak of monkeypox—a relative of smallpox that likely was brought into this country by pet Gambian rats and spread via pet prairie dogs—is only the latest example of the multiple dangers that the importation of exotic pets poses to both animals and humans." "Get a Dog." *Washington Post*, June 13, 2003. Editorial, A28.

[14] See the statement on "Current Visa Restrictions Interfere with U.S. Science and Engineering Contributions to Important National Needs," December 13, 2002, by the National Academies presidents. (Revised June 13, 2003). Available at http://www4.nationalacademies.org/news.nsf/isbn/s12132002?OpenDocument.

[15] Tucker, J.B. 2003. "Preventing the misuse of pathogens: The need for global biosecurity standards," *Arms Control Today*, June: 3-10.

[16] Further information about ICSU may be found at http://www. icsu.org/ while further information about the IAP and the IAC may be found at http://www4.nas.edu/iap/iaphome.nsf?opendatabase and http://www.interacademy council.net, respectively.

Appendix A

ACRONYMS

ACGM	Advisory Committee on Genetic Modification
AEA	Atomic Energy Act
AGSAG	Advisory Group on Scientific Advances in Genetics
AIDS	Acquired Immunodeficiency Syndrome
APHIS	Animal and Plant Health Inspection Service
ASM	American Society for Microbiology
ATCSA	Antiterrorism, Crime and Security Act 2001
BL	Biosafety Level
BMBL	Biosafety in Microbiological and Biomedical Laboratories
BSC	Biological Safety Cabinet
BSO	Biological Safety Officer
BWC	Biological and Toxin Weapons Convention
BW	Biological Weapons
CBW	chemical and biological weapons
CDC	The Centers for Disease Control and Prevention
CIA	Central Intelligence Agency
DARPA	Defense Advanced Research Projects Agency
DHHS	Department of Health and Human Services
DHS	Department of Homeland Security
DOD	Department of Defense

DOE	Department of Energy
DTRA	Defense Threat Reduction Agency
EAR	Export Administration Regulation
EPA	Environmental Protection Agency
FBI	Federal Bureau of Investigation
FDA	Food and Drug Administration
FOIA	Freedom of Information Act
GLP	Good Laboratory Practice
GMO	genetically modified organisms
HEU	Highly Enriched Uranium
HSE	Health and Safety Executive
HWSA	Health and Safety at Work Act of 1974
IAC	InterAcademy Council
IAEA	International Atomic Energy Agency
IAP	InterAcademy Panel on International Issues
IATA	International Air Transport Association
IBC	Institutional Biosafety Committee
ICSU	International Council for Science
IPASS	Interagency Panel for Advanced Science and Security
IRB	Institutional Review Boards
ITAR	International Traffic in Arms Regulations
NATO	North Atlantic Treaty Organization
NHS	National Health Service
NIH	National Institutes of Health
NPT	Nuclear Nonproliferation Treaty
NRC	Nuclear Regulatory Commission
NSA	National Security Agency
NSABB	National Science Advisory Board for Biodefense
NSDD	National Security Decision Directive
NSF	National Science Foundation
OBA	Office of Biotechnology Activities
OECD	Organization for Economic Cooperation and Development
OSTP	Office of Science and Technology Policy
PI	principal investigator
PNAS	*Proceedings of the National Academy of Sciences*
PPE	Personal Protective Equipment

RAC	Recombinant DNA Advisory Committee
RD	Restricted Data
rDNA	Recombinant Deoxyribonucleic Acid
RNA	Ribonucleic Acid

SARS	Severe Acute Respiratory Syndrome
SBIR	Small Business Innovation Research
SBU	Sensitive but unclassified
SEVIS	Student and Exchange Visitor Information System
SIPRI	Stockholm International Peace Research Institute
SPICE	smallpox inhibitor of complement enzymes

UCNI	unclassified controlled nuclear information
UN	United Nations
UNESCO	United Nations Educational, Scientific and Cultural Organization
UPU	Universal Postal Union
USAMRIID	Research Institute for Infectious Diseases
USDA	United States Department of Agriculture

| VCP | *vaccinia* virus complement control protein |

| WHO | World Health Organization |
| WMD | Weapons of Mass Destruction |

Appendix B

Biographical Sketches of Committee Members

Gerald R. Fink (NAS, IOM) is a professor of genetics, Whitehead Institute, Massachusetts Institute of Technology. Dr. Fink is a founding member of the Whitehead Institute and American Cancer Society Professor of Genetics at MIT, Dr. Fink was Director of the Whitehead Institute from 1990 to 2001. He received his B.A. from Amherst College in 1962 and his Ph.D. from Yale in 1965. In addition, he has received honorary doctorates from Amherst College and Cold Spring Harbor Laboratory. His research focuses on the molecular biology of fungal infectious disease. He served as president of the Genetics Society of America. Among his many awards are the National Academy of Sciences Award in Molecular Biology, the Medal of the Genetics Society of America, Emil Christian Hansen Award (Denmark), the Yale Science and Engineering Award, and the 2001 George Beadle Award. He has been elected to the National Academy of Sciences, the American Academy of Arts and Sciences, the Institute of Medicine and the American Philosophical Society. He is currently a Senior Scholar in Infectious Disease of the Ellison Foundation.

Ronald Atlas is a professor of biology and graduate dean at the University of Louisville. He is the Director of the Center for Deterrence of Biowarfare and Bioterrorism. Dr. Atlas' studies have focused on the application of molecular techniques to environmental problems. His studies have included the development of "suicide vectors" for the containment of genetically engineered microorganisms and the use of gene probes and the polymerase chain reaction for environmental monitoring. He received a BS degree from the State University of New York at Stony Brook in 1968, an M.S. from Rutgers University in 1970 and a Ph.D. from Rutgers University in 1972. He then served for a year as a National Research Council Research Associate at the Jet Propulsion Laboratory. He is the immediate past-president of the American Society for Microbiology and was the recipient of the American Society for Microbiology award in Applied and Environmental Sciences.

W. Emmett Barkley is Director of Laboratory Safety at the Howard Hughes Medical Institute (HHMI). Dr. Barkley directed the National Cancer Institute's Office of Research Safety and the divisions of safety and engineering services at the National Institutes of Health prior to joining HHMI. He received his B.S. in civil engineering from the University of Virginia and his M.S. and Ph.D. in environmental health from the University of Minnesota. Dr. Barkley has received several awards including the Distinguished Service Medal of the U.S. Public Health Service. He has previously served the Academies on several committees including the NRC Committee on Prudent Practices for Handling, Storage, and Disposal of Chemicals in the Laboratory and served as the chair of the Committee on Safety and Health in Research Animal Facilities.

R. John Collier (NAS) is Presley Professor of Microbiology and Molecular Genetics in the Department of Microbiology and Molecular Genetics at Harvard Medical School. His scientific contributions include: demonstrating that Diphtheria toxin blocks protein synthesis by inactivating Elongation Factor-2; elucidating Diphtheria toxin structure, as containing enzymatic (A) and binding (B) fragments; identifying GLU-148 as a key active-site residue; developing A-chain immunotoxin concepts; crystallizing Diphtheria and Pseudomonas toxins and determining Pseudomonas toxin's structure. His studies in recent years have focused on the structure and mode of action of anthrax toxin, and ways to inhibit its action.

Susan E. Cozzens is Professor and Chair of the School of Public Policy at the Georgia Institute of Technology. Her current research is on science, technology, and inequalities. She is active internationally in developing research assessment methods and science and technology indicators. Dr. Cozzens had previously been Director of the Office of Policy Support at the National Science Foundation and has served as a consultant to several organizations including the Committee on Science, Engineering and Public Policy of the National Research Council, Office of Science and Technology Policy, National Science Foundation, Institute of Medicine, Office of Technology Assessment, General Accounting Office, National Cancer Institute, National Institute on Aging, National Institutes of Health, and the National Institute on Occupational Safety and Health. She has given speeches on science policy and research evaluation all around the world and has many publications on science policy and science and technology studies.

Ruth Faden (IOM) is Philip Franklin Wagley Professor of Biomedical Ethics and Executive Director of the Phoebe R. Berman Bioethics Institute, Johns Hopkins University. Dr. Faden is also a Senior Research Scholar at the Kennedy Institute of Ethics, Georgetown University. She is a Fellow of the Hastings Center and the American Psychological Association and has served on several national advisory committees and commissions.

David R. Franz is currently Vice President of Chemical & Biological Defense Division at Southern Research Institute. He has served in the U.S. Army Medical Research and Materiel Command for 23 of his 27 years on active duty. Dr. Franz has served as both Deputy Commander and then Commander of the U.S. Army Medical Research Institute of Infectious Diseases (USAMRIID) and as Deputy Commander of the U.S. Army Medical Research and Materiel Command. Prior to joining the Command, he served as Group Veterinarian for the 10th Special Forces Group (Airborne). Dr. Franz served as Chief Inspector on three United Nations Special Commission biological warfare inspection missions to Iraq, and as technical advisor on long-term monitoring. He also served as a member of the first two US/UK teams that visited Russia in support of the Trilateral Joint Statement on Biological Weapons, and as a member of the Trilateral Experts Committee for biological weapons negotiations. While at the Medical Research and Materiel Command, he was assigned to four of its laboratories, personally conducting research and publishing in the areas of frostbite pathogenesis, organophosphate chemical warfare agent effects on pulmonary and upper airways function, the role of cell-mediated small vessel dysfunction in cerebral malaria, and the development of medical countermeasures against biological agents. Dr. Franz was Technical Editor for the Textbook of Military Medicine on Chemical and Biological Defense released in 1997.

Joseph L. Goldstein (NAS, IOM) is Regental Professor and Chairman of the Department of Molecular Genetics and Paul J. Thomas Professor of Medicine and Genetics at the University of Texas Southwestern Medical Center at Dallas. Together with his colleague Dr. Michael S. Brown, Dr. Goldstein has received numerous awards — including the Albert D. Lasker Award in Basic Medical Research, the Nobel Prize in Physiology or Medicine, and the National Medal of Science — for their discovery of receptors that control cholesterol metabolism, Dr. Goldstein's area of expertise is human genetics and cholesterol metabolism. He is a member of the U.S. National Academy of Sciences, American Philosophical Society, and the Institute of Medicine. He is also a Foreign Member of The Royal Society (London). He is a past president of the American Society for Clinical Investigation and was a member of the Governing Council of the U.S. National Academy of Sciences. He has also served on the Scientific Review Board and Medical Advisory Board of the Howard Hughes Medical Institute and on the NIH Program Advisory Committee on the Human Genome. He was a Non-Resident Fellow of The Salk Institute. Dr. Goldstein is currently Chairman of the Medical Advisory Board of the Howard Hughes Medical Institute, Chairman of the Albert Lasker Medical Research Awards Jury, and a member of the Board of Trustees of The Rockefeller University. He is a member of the Scientific Advisory Boards of the Welch Foundation, Memorial-Sloan Kettering Medical Center, and the Scripps Research Institute.

Robert Kadlec is a Professor of Military Strategy and Operations at National Defense University. He is an Air Force physician who joined the National War College Faculty in December 1999. Col. Kadlec served as a Senior Assistant for Counter proliferation Policy in the Office of the Secretary of Defense (OSD) for Policy, was a member of the Joint Chiefs of Staff, and represented OSD on the U.S. delegation to the Biological Weapons Convention in Geneva, Switzerland. He is board certified in both General Preventive Medicine & Public Health and Aerospace Medicine and is an Assistant Clinical Professor of Military Medicine at USUHS. Col. Kadlec also served as an UNSCOM inspector in Iraq.

Barry Kellman is the Director of the International Weapons Control Center at the DePaul University College of Law. He served as legal adviser to the National Commission on Terrorism in 2000, and is currently a consultant for the Memorial Institute for the Prevention of Terrorism, preparing a series of monographs on Legal Authorities and Liabilities for Catastrophic Terrorism. He chairs the ABA Committee on Law and National Security as well as the Arms Control Section of the American Society of International Law. In addition to his work on terrorism, Professor Kellman is a legal authority on the Chemical Weapons Convention and has served as a consultant to the Defense Department on a wide array of weapons control issues. Since 1995, he has participated in Track-2 discussions of Middle East arms control. He has published widely on weapons proliferation and smuggling, national security, and the laws of armed conflict; and he has written the only legal publications on biological terrorism: Biological Terrorism: Legal Measures for Preventing Catastrophe, Spring 2001; and An International Criminal Law Approach To Bio-Terrorism, Spring 2002.

Marc Kirschner (NAS) is Chair and Carl W. Walter Professor of Cell Biology, Department of Cell Biology, Harvard Medical School. His main areas of study are cell biology, cytoskeleton, cell cycle, and vertebrate embryology. Dr. Kirschner is well known for discoveries on microtubule assembly and the analysis of tubulin genes and for the contributions to molecular analysis of amphibian development, especially the control of the early cell cycles during embryogenesis and molecular event in embryonic induction. Dr. Kirschner was elected Foreign member of the Royal Society of London in 1999. He is a member of the American Academy of Arts and Sciences and has served on the Advisory Committee to the Director of the National Institutes of Health and as President of the American Society for Cell Biology.

Erin O'Shea is a Professor and Vice Chair of the Department of Biochemistry & Biophysics at the University of California at San Francisco and an Assistant Investigator of the Howard Hughes Medical Institute. Her research interests include signal transduction and gene regulation and the use of genomic and proteomic approaches to study eukaryotic cells. Her awards include a David and Lucile Packard Fellowship, a Presidential Faculty Fellow Award, the American Society for Cell Biology-Promega Early Career Life Scientist Award, and the National Academy of Sciences Award in Molecular Biology. She has served in several advisory roles, including: Scientific Advisory Board, Helen Hay Whitney Foundation; Chairman, Scientific Advisory Board, Boston University School of Medicine Department of Genetics and Genomics; External Review Committee, Lawrence Berkeley National Laboratory Division of Physical Biosciences. She has previously served the Academies on the HHMI Predoctoral Fellowships Panel on Biochemistry and Structural Biology.

Clarence J. Peters is currently a professor in the Department of Microbiology & Immunology and Pathology at the University of Texas Medical Branch in Galveston. He had been Chief of Special Pathogens at the Centers for Disease Control in Atlanta, Georgia. Formerly Chief of the Disease Assessment Division at the U.S. Army Medical Research Institute of Infectious Diseases (USAMRIID), he has worked in the field of infectious diseases for three decades with the CDC, the U.S. Army, and the U.S. Public Health Service. He was the head of the unit that contained the outbreak of Ebola in Rhesus macaques at a Reston, Virginia animal facility. He was also called in to contain an outbreak of deadly hemorrhagic fever in Bolivia. He received his M.D. from Johns Hopkins University and has more than 275 publications in the areas of virology and viral immunology. Dr. Peters is currently a member of the National Research Council Committee on Occupational Health and Safety in Care of

Nonhuman Primates and the Committee on Emerging Microbial Threats to Health in the 21st Century.

Judith V. Reppy is a professor in the Department of Science and Technology Studies and associate director of the Peace Studies Program of Cornell University. She is an adjunct member of the Department of Government. She has been a visiting fellow at Science & Technology Studies (Manchester University), the Science Policy Research Unit (Sussex University), and the Center for International Studies (MIT). She is a member of the Council on Foreign Relations, the Boards of Directors of The Federation of American Scientists, Economists Allied for Arms Reduction (ECAAR) and the Institute for Defense and Disarmament Studies (IDDS), and the Advisory Board of Women in International Security (WIIS). She served as co-chair of US Pugwash from 1995-2000. Dr. Reppy is the author, co-author, and contributing editor of several books, as well as numerous articles and contributed chapters in edited works.

Elizabeth Rindskopf Parker is Dean of the University of the Pacific McGeorge School of Law. Ms. Rindskopf Parker is a leading expert on anti-terrorism law. Her expertise includes law of national security and terrorism; international relations; public policy and technology development; and transfer, commerce, and litigation in the area of civil rights and liberties. Ms. Rindskopf Parker was General Counsel to the University of Wisconsin System and Counsel to the international law firm of Bryan Cave, LLP where her practice focused on counseling clients on public policy and international trade issues. She previously served as the General Counsel for the Central Intelligence Agency; Principal Deputy Legal Adviser for the U.S. Department of State; General Counsel for the National Security Agency, Department of Defense; and Acting Assistant Director for the Federal Trade Commission. Ms. Rindskopf Parker often speaks on subjects dealing with the law of national security. She is a member of the Council on Foreign Relations, past chair of the ABA Standing Committee on Law and National Security and currently a member of the ABA President's Task Force on the Laws of Terrorism.

Matthew Scharff (NAS) is Professor of Cell Biology, Albert Einstein School of Medicine. Dr. Scharff has served in several advisory roles including, Outside Advisory Committees for NIEHS Center of Environmental Medicine at N.Y.U. Cancer Center at N.Y.U. and the Cancer Center, University of Pennsylvania, Scientific Advisory Board, Helen Hay Whitney Foundation; chairman, Scientific Advisory Board, Rappaport Family Institute, Haifa, Israel; Scientific Advisory Board, City of Hope; Scientific Advisory Board, Simons Arthritis Center, University of Texas,

Southwestern Medical School, Dallas; Co-Chairman, Board of Scientific Counselors, Division of Basic Sciences, NCI; member of NCI Executive Committee; Advisory Committee to the Director of the NCI. His current research is being used to create better monoclonal antibodies for the treatment and prevention of disease.

Morton Schwartz is Chairman of the Department of Clinical Laboratories and Head of Applied and Diagnostic Biochemistry at the Memorial Sloan Kettering Cancer Center. After serving with the U.S. Navy during World War II, he was a U.S. Public Health Service Fellow from 1950-1952. Dr. Schwartz is a member of several professional associations , including the American Association of Clinical Chemistry, which gave him its Service to the Profession Award in 1988. His expertise is Clinical Chemistry.

Edward Scolnick (NAS, IOM) is President Emeritus, Merck Research Laboratories and Executive Vice President for Science and Technology with Merck & Company, Inc. Dr. Scolnick was elected to the National Academy of Sciences in 1984 and to the American Academy of Arts and Sciences in 1993. He became a member of the Institute of Medicine in 1996 and in 1997 was elected to the Merck & Co., Inc. Board of Directors. He currently serves on the Board of Directors for Millipore Corporation, Renovis, Inc., Cold Spring Harbor Laboratory, Harvard Medical School, Protein Pathways, GeneSoft, Inc., TransForm Pharmaceuticals, Inc., the Medical and Scientific Advisory Board for MPM Capital, and was recently appointed to the Governor's Pennsylvania Health Research Advisory Committee. In addition, he is a Member of the FDA Science Board. Dr. Scolnick's commitment to the mental health field is evidenced by memberships on the Board of Directors for McLean Hospital, McGovern Institute for Brain Research, Pennsylvania Montgomery County Emergency Services, and as President of the Pennsylvania Montgomery County Chapter of the National Alliance for the Mentally Ill. He is also a board member of the National Institute of Mental Health Council.

Appendix C

Committee Meetings

First Meeting
April 1-2, 2002
Washington, D.C.

Meeting Objectives: Introduce National Research Council procedures; committee introductions and composition/balance/bias discussions; committee and report procedures; discuss genesis of the study and Statement of Task; discuss draft report outline; discuss project plan and report realization; receive overview briefing on the current U.S. regulatory environment; determine objectives and date of next committee meeting.

Presenters

Analysis of the current (U.S.) regulatory environment: "Select Agent Rule," RAC/IRB Rules and Practices, Effectiveness/Enforcement of Current (U.S.) Biotechnology Rules and Practices

Joseph G. Perpich
Perpich and Associates, Inc.

Ronald Atlas
American Society for Microbiology

The view from "The Hill" – Congressional perspectives on potentially dangerous biotechnology research and pathogens

Stephen Redhead
Congressional Research Service

View from the Executive Branch: "Safeguarding Information Regarding Weapons of Mass Destruction" – Office of Homeland Security, Executive Office of the President

Penrose (Parney) C. Albright
Office of Homeland Security
Office of Science and Technology Policy

Rachel Levinson
Office of Science and Technology Policy

"Protective Oversight of Biotechnology: A Discussion Paper"

John Steinbruner
University of Maryland

Second Meeting
June 24-25, 2002
Washington, D.C.

Meeting Objectives: Introduce new members and complete composition/balance/bias discussion; discuss draft report outline; receive briefings on defining the problem, safeguarding information and governmental challenges; make writing assignments; determine objectives and date of next committee meeting.

Presenters

Defining the Universe of Potentially Dangerous Biotechnology Research

Gerald Epstein
Defense Threat Reduction Agency

Clarence J. Peters, Professor
University of Texas at Galveston

Mark Wheelis, Professor
University of California at Davis

Safeguarding Information in the Life Sciences

Steven M. Block, Professor
Stanford University

Eugene B. Skolnikoff, Professor
Massachusetts Institute of Technology

The Life Sciences Community and the Safeguarding of Scientific Knowledge: Challenges for Government

Guy Roberts
Department of the Navy

R. Timothy Mulcahy, Associate Dean and Professor
University of Wisconsin, Madison

Steven Aftergood, Senior Research Analyst
Federation of American Scientists

Third Meeting
September 9-10, 2002
Washington, D.C.

Meeting Objectives: Receive briefings on "science and security issues" and the current thinking on "information security"; discuss chapter drafts; make writing assignments; determine objectives and date of next committee meeting.

Presenters

Defining Potentially Dangerous Research in the Life Sciences

Malcolm Dando
University of Bradford, U.K.

Defining "Sensitive" Information in the Life Sciences

Parney Albright
Office of Science and Technology Policy

Rachel Levinson
Office of Science and Technology Policy

Defining "Sensitive" Information in the Life Sciences—The NIH Perspective

Anthony Fauci, Director
National Institute of Allergy and Infectious Disease
National Institutes of Health

"Life Sciences Research in International Security: Policy Developments for Responsible Use of 'Dual-Use' Knowledge"

George Poste
Health Technology Networks

Biological Weapons Working Group U.K. Consultation Overview

John Steinbruner
University of Maryland

Fourth Meeting
October 8, 2002
Washington, D.C.

Meeting Objectives: Review of contents, structure and substance of draft chapters; discuss report plan and status; make writing assignments; determine objectives and date of next committee meeting.

No Presenters

Fifth Meeting
November 11, 2002
Washington, D.C.

Meeting Objectives: Review of contents, structure and substance of draft chapters; discuss report plan and status; make writing assignments; determine objectives and date of next committee meeting.

No Presenters

Sixth Meeting
January 29, 2003
Washington, D.C.

Meeting Objectives: Review of contents, structure and substance of draft chapters; discuss report plan and status; make writing assignments.

No Presenters